SpringerBriefs in Mathematical Physics

Volume 52

SpringerBriefs are characterized in general by their size (50–125 pages) and fast production time (2–3 months compared to 6 months for a monograph).
Briefs are available in print but are intended as a primarily electronic publication to be included in Springer's e-book package.

Typical works might include:

- An extended survey of a field
- A link between new research papers published in journal articles
- A presentation of core concepts that doctoral students must understand in order to make independent contributions
- Lecture notes making a specialist topic accessible for non-specialist readers.

SpringerBriefs in Mathematical Physics showcase, in a compact format, topics of current relevance in the field of mathematical physics. Published titles will encompass all areas of theoretical and mathematical physics. This series is intended for mathematicians, physicists, and other scientists, as well as doctoral students in related areas.

Sampei Hirose • Naofumi Honda •
Takahiro Kawai • Shinji Sasaki • Yoshitsugu Takei

Virtual Turning Points II

Their Interplay with Integral Representations and Non-Hereditary Turning Points

Springer

Sampei Hirose
Department of Engineering and Design
Shibaura Institute of Technology
Saitama, Japan

Naofumi Honda
Department of Mathematics
Hokkaido University
Sapporo, Hokkaido, Japan

Takahiro Kawai
RIMS
Kyoto University
Kyoto, Kyoto, Japan

Shinji Sasaki
OCAMI
Osaka Metropolitan University
Osaka, Japan

Yoshitsugu Takei
Department of Mathematical Sciences
Doshisha University
Kyotanabe, Kyoto, Japan

ISSN 2197-1757 ISSN 2197-1765 (electronic)
SpringerBriefs in Mathematical Physics
ISBN 978-981-95-2816-5 ISBN 978-981-95-2817-2 (eBook)
https://doi.org/10.1007/978-981-95-2817-2

This Springer imprint is published by the registered company Springer Nature Singapore Pte Ltd.
The registered company address is: 152 Beach Road, #21-01/04 Gateway East, Singapore 189721, Singapore

In memory of the late Professor Tatsuya Koike

Preface

This is the sequel to *Virtual Turning Points* (Springer Briefs in Mathematical Physics, Vol. 4, referred to as [VTP]) written by N. Honda, T. Kawai, and Y. Takei, three of the authors of the present volume. As we noted in the Preface to [VTP], global asymptotic analysis of higher-order differential equations is a very difficult and troublesome subject (cf. [F]). To understand this subject, Aoki et al. [AKT1] introduced the notion of virtual turning points by applying the technique of microlocal analysis. Virtual turning points and new Stokes curves, i.e., Stokes curves emanating from virtual turning points which were originally introduced by Berk, Nevins, and Roberts in [BNR] without using virtual turning points, have renewed the appearance of asymptotic analysis of higher-order equations, but still complete understanding of the subject has not been achieved yet. In fact, since the publication of [VTP], several important and interesting phenomena have been newly discovered—for example, non-hereditary turning points, crossing of three ordinary Stokes curves, leaf-type and tadpole-type configurations of Stokes curves, and so on. The aim of this second volume is to explain these new phenomena related to virtual turning points and to study their intriguing properties mainly by examining concrete examples of differential equations with integral representations of solutions.

To be more concrete, in Chap. 1, after reviewing the most fundamental parts of [VTP], we introduce the notion of non-hereditary turning points, which are the main subject of this volume and naturally appear when we consider a tangential system. A non-hereditary turning point produces no ill-effect as it is, but it creates a new class of virtual turning points and an additional period associated with them if we perturb the equation while keeping its principal part intact. We present some illuminating and inspiring examples that are concerned with the effect of a non-hereditary turning point and a crossing point of three ordinary Stokes curves in Sect. 1.5, which will be discussed in a more detailed manner later in Chap. 3. Then, in Chap. 2, we try to answer the question of how these new phenomena such as crossing of three ordinary Stokes curves are related to the location of singularities of the Borel transformed WKB (Wentzel–Kramers–Brillouin) solutions. In particular, we study these relationships from the viewpoint of the periods and the "growing trees" discussed in [H] and [VTP, Chap. 2]. Taking the argument of Chap. 2 into account, we clarify that a new class of virtual turning points and an additional period created by a non-hereditary turning point originate from the so-called "bicharacteristic chain" ([VTP, p. 84]) of the Borel transform of the relevant system in Chap. 3. We also compute in Sect. 3.4 the explicit form of connection formulas for WKB solutions concerned with the leaf-type and tadpole-type configurations of Stokes curves. Although newly discovered intriguing notions and phenomena are investigated mainly by using concrete examples in this second volume, we believe these new objects such as non-hereditary turning points are significant and will play an important role in asymptotic analysis of higher-order differential equations together with virtual turning points themselves in the future.

As mentioned at the end of Sect. 1.2, it is the late Professor Tatsuya Koike who suggested to us the importance of non-hereditary turning points in considering the perturbation of tangential systems. We deeply appreciate his contribution to the subject and would like to dedicate this volume to him. We sincerely thank also Prof. T. Aoki, with whom we have had many fruitful discussions since the discovery of the virtual turning point. Further, we are grateful to Prof. S. Kamimoto and Prof. K. Iwaki for their collaboration with (some of) us. Last but not least, we sincerely thank M. Nakamura of Springer Japan for his kind support in publishing this volume in addition to [VTP].

Saitama, Japan — Sampei Hirose
Sapporo, Japan — Naofumi Honda
Kyoto, Japan — Takahiro Kawai
Osaka, Japan — Shinji Sasaki
Kyotanabe, Japan — Yoshitsugu Takei
March 2025

Acknowledgments S. Hirose is supported by JSPS KAKENHI Grant Number 21K03300, N. Honda is supported by JSPS KAKENHI Grant Number 21K03284, S. Sasaki is partly supported by JSPS KAKENHI Grant Number JP21K13811, and Y. Takei is supported by JSPS KAKENHI Grant Number 19H01794 and 24K06767. The research related to this volume was supported also by the Research Institute for Mathematical Sciences, an International Joint Usage/Research Center located in Kyoto University.

Contents

Chapter 1
Integral Representation of Solutions and Related Topics

Abstract As explained in the first volume (Honda et al. Virtual Turning Points. Springer Briefs in Mathematical Physics, vol. 4. Springer-Verlag, Berlin, 2015), the virtual turning points are defined in terms of the bicharacteristic strips of (the Borel transform of) a system of differential equations. However, when we consider its tangential system, a new kind of turning points called non-hereditary turning points appears. Using a system with an integral representation of solutions mainly, we discuss non-hereditary turning points in this chapter. After reviewing the definition of a virtual turning point and an s-virtual turning point, its special version for a system with integral representations, we explain the non-hereditary turning points and their fundamental properties. As an example, we discuss a tangential system of the Pearcey system to a hyperplane in detail. We also present some intriguing examples of the Stokes geometry of systems with integral representation of solutions.

1.1 A Virtual Turning Point and an s-Virtual Turning Point

Following [VTP], we first recall the definition of virtual turning points and s-virtual turning points and briefly review the background for them in this section.

We consider an ordinary differential equation with a large parameter η,

$$P\psi = \left(\frac{d^m}{dx^m} + q_1(x)\eta\frac{d^{m-1}}{dx^{m-1}} + \cdots + q_m(x)\eta^m\right)\psi = 0. \tag{1.1.1}$$

Here and in what follows we assume the large parameter η is a positive number and $q_j(x)$ $(j = 1, 2, \ldots, m)$ is a polynomial or a rational function. A solution of (1.1.1) of the form

$$\psi(x, \eta) = \eta^{-1/2}\exp\left(\int_{x_*}^{x} S(x, \eta)dx\right) \tag{1.1.2}$$

S. Hirose et al., *Virtual Turning Points II*, SpringerBriefs in Mathematical Physics 52, https://doi.org/10.1007/978-981-95-2817-2_1

with

$$S(x, \eta) = \eta S_{-1}(x) + S_0(x) + \eta^{-1} S_1(x) + \cdots \tag{1.1.3}$$

is called a **WKB solution** of (1.1.1). Here x_* is an appropriately fixed point and $S(x, \eta)$ satisfies a non-linear differential equation of order $(m-1)$ called the **higher-order version of the Riccati equation**. See [KT] and [VTP, Sections 1.1 and 1.2] for the details of the construction of WKB solutions. In particular, the top-order term $S_{-1}(x)$ of $S(x, \eta)$ satisfies

$$\big(S_{-1}(x)\big)^m + q_1(x)\big(S_{-1}(x)\big)^{m-1} + \cdots + q_m(x) = 0, \tag{1.1.4}$$

that is, $S_{-1}(x)$ is a root of

$$\tilde{p}(x, \zeta) = \zeta^m + q_1(x)\zeta^{m-1} + \cdots + q_m(x) = 0. \tag{1.1.5}$$

If we let $\zeta_j(x)$ $(j = 1, 2, \ldots, m)$ denote the solutions of (1.1.5), there exist m linearly independent WKB solutions $\psi_j(x, \eta)$ of (1.1.1) with the choice of $S_{-1}(x) = \zeta_j(x)$, which can be rewritten also as

$$\exp\left(\eta \int_{x_*}^{x} \zeta_j(x)dx\right) \sum_{n=0}^{\infty} \psi_{j,n}(x)\eta^{-(n+1/2)}. \tag{1.1.6}$$

In the following discussion we often deal with the WKB solutions $\psi_j(x, \eta)$ $(j = 1, 2, \ldots, m)$ after introducing appropriate cuts in the x-plane so that $\zeta_j(x)$ may be single-valued on the cut plane.

Using the single-valued solutions $\zeta_j(x)$ thus defined, we introduce the following terminologies which will be frequently used in this volume.

Definition 1.1

(i) A point $x = a$ is said to be a **turning point** of (1.1.1) if (1.1.5) has a multiple solution in ζ there. In what follows, except for some exceptional cases, we assume only two solutions $\zeta_j(x)$ and $\zeta_k(x)$ $(j \neq k)$ coalesce, that is,

$$\zeta_j(a) = \zeta_k(a) \quad (j \neq k) \tag{1.1.7}$$

holds at a turning point. When (1.1.7) holds, the turning point $x = a$ is said to be **of type** (j, k).

(ii) Let $x = a$ be a turning point of type (j, k). Then a **Stokes curve of type** (j, k) that emanates from a is, by definition, the curve given by

$$\operatorname{Im} \int_a^x \big(\zeta_j(x) - \zeta_k(x)\big)dx = 0. \tag{1.1.8}$$

Definition 1.2 Let $x = a$ be a turning point of type (j, k). Then each segment of the Stokes curve emanating from $x = a$ is **labeled as** $(j > k)$ **or** $(j < k)$, according to whether

$$\mathrm{Re}\int_a^x \big(\zeta_j(x) - \zeta_k(x)\big)dx > 0 \tag{1.1.9}$$

or

$$\mathrm{Re}\int_a^x \big(\zeta_j(x) - \zeta_k(x)\big)dx < 0 \tag{1.1.10}$$

holds on the segment.

Definition 1.3 If exactly two solutions $\zeta_j(x)$ and $\zeta_k(x)$ $(j \neq k)$ of (1.1.5) coalesce at $x = a$, and if

$$\left.\frac{\partial \tilde{p}}{\partial x}\right|_{x=a,\zeta=\zeta_j(a)} \neq 0, \tag{1.1.11}$$

then the point $x = a$ is said to be a **simple turning point**.

Remark 1.1 By its definition we find a turning point is a zero of the discriminant of (1.1.5). We sometimes refer to the turning point in Definition 1.1 as the "**ordinary turning point**" to distinguish it from the virtual turning point. Similarly, the Stokes curve in Definition 1.1 is sometimes referred to as the "**ordinary Stokes curve**".

WKB solutions are divergent in general and we give them an analytic interpretation through the **Borel resummation method** in the exact WKB analysis (cf. [KT], [V]). That is, we consider the **Borel sum** of $\psi_j(x, \eta)$ defined by

$$\int_{\gamma_j} \exp(-y\eta)\psi_{j,B}(x, y)dy, \tag{1.1.12}$$

where

$$\psi_{j,B}(x, y) = \sum_{n=0}^{\infty} \frac{\psi_{j,n}(x)}{\Gamma(n + 1/2)} \left(y + \int_{x_*}^x \zeta_j(x)dx\right)^{n-1/2} \tag{1.1.13}$$

is the **Borel transform** of $\psi_j(x, \eta)$ and the integration path γ_j is given by

$$\gamma_j = \left\{y \in \mathbb{C}\,;\ \mathrm{Im}\left(y + \int_{x_*}^x \zeta_j(x)dx\right) = 0,\ \mathrm{Re}\left(y + \int_{x_*}^x \zeta_j(x)dx\right) \geq 0\right\}. \tag{1.1.14}$$

A WKB solution $\psi_j(x,\eta)$ is said to be **Borel summable** if its Borel transform $\psi_{j,B}(x,y)$ converges in a neighborhood of $y = -\int_{x_*}^{x} \zeta_j(x)dx$ and can be analytically continued along γ_j given by (1.1.14), and further if the integral (1.1.12) is well-defined. Note that, in defining the Borel sum (1.1.12), we fix the variable x and regard it just as a parameter. In this analysis the most important is the singularity structure of the Borel transformed WKB solution $\psi_{j,B}(x,y)$. For example, let a be a simple turning point of type (j,k) in the sense of Definitions 1.1 and 1.3. Then, as is explained in [VTP, Section 1.2], the Borel transform $\psi_{j,B}(x,y)$ of a WKB solution $\psi_j(x,\eta)$ suitably normalized at $x = a$ has singularities at

$$\Gamma_j = \{(x,y)\ ;\ y = y_j(x)\} \text{ and } \Gamma_k = \{(x,y)\ ;\ y = y_k(x)\}, \tag{1.1.15}$$

where $y_j(x) = -\int_a^x \zeta_j(x)dx$ and $y_k(x) = -\int_a^x \zeta_k(x)dx$, in a neighborhood of $(x,y) = (a,0)$. (The singularity structure of $\psi_{k,B}(x,y)$ is also the same as $\psi_{j,B}(x,y)$ near $(a,0)$.) This singularity structure of $\psi_{j,B}(x,y)$ implies that a **Stokes phenomenon** $\psi_j \to \psi_j \pm i\psi_k$ occurs with the Borel sum (1.1.12) of $\psi_j(x,\eta)$ when the singular point $y_k(x)$ of $\psi_{j,B}(x,y)$ hits the integration path (1.1.14) (with x_* being replaced by a), that is, when x lies on a Stokes curve of type $(j > k)$, under the assumption that $\psi_{j,B}(x,y)$ can be analytically continued endlessly without encountering the natural boundaries and further that its growth order near $y = \infty$ is tame. (Note that these assumptions are confirmed for second-order differential equations in [KoS], but it is still an open problem to verify these assumptions for higher-order equations.) Thus the Stokes phenomena for WKB solutions of (1.1.1) are induced by the interplay of two "cognate" singularities of the Borel transformed WKB solutions that merge at a turning point.

A turning point plays an important role in the exact WKB analysis as two cognate singularities of a Borel transformed WKB solution merge and consequently Stokes curves emanate from there. However, in the case of a higher-order (i.e., $m \geq 3$) differential Eq. (1.1.1), a point called a virtual turning point also plays a similar role to that of a turning point. A key in defining a virtual turning point is the following two basic facts:

(I) The Borel transform of a WKB solution satisfies a partial differential equation $P_B\psi_B = 0$, where

$$P_B = P_B\Big(x, \frac{\partial}{\partial x}, \frac{\partial}{\partial y}\Big) = \frac{\partial^m}{\partial x^m} + q_1(x)\frac{\partial^{m-1}}{\partial x^{m-1}}\frac{\partial}{\partial y} + \cdots + q_m(x)\frac{\partial^m}{\partial y^m} \tag{1.1.16}$$

is the Borel transform of P.

(II) The singularities of solutions of $P_B\psi_B = 0$ propagate along a bicharacteristic strip if the operator P_B is with **simple characteristics**, that is, a bicharacteristic strip is the most elementary carrier of singularities (cf. [SKK]; note that P_B is with simple characteristics when all turning points are simple). Here a **bicharacteristic strip** associated with P_B is, by definition, a solution curve

$(x(t), y(t); \xi(t), \eta(t))$ in the cotangent bundle $T^*\mathbb{C}^2_{(x,y)}$ of the Hamilton–Jacobi equation

$$\frac{dx}{dt} = \frac{\partial\sigma}{\partial\xi}, \quad \frac{dy}{dt} = \frac{\partial\sigma}{\partial\eta}, \quad \frac{d\xi}{dt} = -\frac{\partial\sigma}{\partial x}, \quad \frac{d\eta}{dt} = -\frac{\partial\sigma}{\partial y} \tag{1.1.17}$$

that is contained in the **characteristic variety**

$$V = \{(x, y; \xi, \eta) \in T^*\mathbb{C}^2 \; ; \; \sigma(x, \xi, \eta) = 0, \; \eta \neq 0\} \tag{1.1.18}$$

of P_B, where $\sigma = \sigma(x, \xi, \eta)$ denotes the **(principal) symbol** of $P_B(x, \partial/\partial x, \partial/\partial y)$, i.e., $P_B(x, \xi, \eta)$. (As σ is free from y, (1.1.17) implies η is a constant, which we choose to be 1 later.)

Based on these facts, the definition of a virtual turning point is given as follows:

Definition 1.4 The x-component of a self-intersection point of the **bicharacteristic curve**, that is, the projection of a bicharacteristic strip to the base manifold $\mathbb{C}^2_{(x,y)}$, is called a **virtual turning point** of (1.1.1).

The cognate singularities of a Borel transformed WKB solution merging at a simple turning point are nothing but the two portions of the projection to $\mathbb{C}^2_{(x,y)}$ of one bicharacteristic strip, which is a non-singular curve in $T^*\mathbb{C}^2$. Otherwise stated, a turning point is a confluent point of two cognate singularities, which are two components in the projection of a bicharacteristic strip. Similarly, at a virtual turning point two cognate singularities merge as a virtual turning point is a self-intersection point of a bicharacteristic curve, i.e., the projection of a bicharacteristic strip. Since each component of the bicharacteristic curve is locally expressed as $y = -\int^x \zeta_j(x)dx$ $(j = 1, 2, \ldots, m)$ outside the set of turning points (cf. [VTP, (1.4.21)]), the type of a virtual turning point can be defined similarly to ordinary turning points.

In the study of the Borel resummed WKB solutions of higher-order differential equations virtual turning points and Stokes curves emanating from them are indispensable. For example, the **BNR equation**

$$\left(\frac{d^3}{dx^3} + 3\eta^2\frac{d}{dx} + 2ix\eta^3\right)\psi = 0 \tag{1.1.19}$$

has a unique virtual turning point at $x = 0$ in addition to ordinary turning points $x = \pm 1$, and the Stokes curve emanating from $x = 0$ gives the **new Stokes curves** that Berk et al. [BNR] proposed to introduce. For more details see [VTP, Sections 1.3 and 1.4]. Respecting the importance of the discovery of [BNR], we sometimes refer to a Stokes curve emanating from a virtual turning point as a new Stokes curve. In this volume, however, a Stokes curve normally means a Stokes curve emanating from either an ordinary turning point or a virtual turning point. As is verified in [VTP, Proposition 1.4.2], near a virtual turning point no Stokes phenomena for

WKB solutions are observed on a Stokes curve emanating from it, that is, a Stokes curve is **inert** near a virtual turning point. If a Stokes curve emanating from a virtual turning point is inert at any point in the curve, we say that the virtual turning point is **redundant**. In what follows, an inert portion of a Stokes curve is described by a dotted line. On the other hand, it often occurs that a Stokes curve emanating from a virtual turning point becomes **active** after it passes over an **ordered crossing point**, i.e., a crossing point of two Stokes curves $\mathcal{S}_1$ and $\mathcal{S}_2$ of type (j, k) and (k, l), respectively, with either $j > k > l$ or $j < k < l$ being satisfied. A typical example of the latter situation is the BNR equation (1.1.19). In the case of the BNR equation, two Stokes curves emanating from the ordinary turning points $x = \pm 1$ make an ordered crossing point at $x = C_1$ on the positive imaginary axis and a new Stokes curve $\mathcal{S}_{C_1}$ emanating from the virtual turning point $x = 0$ becomes active after it passes over this ordered crossing point C_1. (The same phenomenon is also observed at an ordered crossing point C_2 on the negative imaginary axis; cf. Fig. 1.1.) In general, if we encounter a crossing point of two Stokes curves, we can locate a virtual turning point from this crossing point conversely (cf. [VTP, Proposition 1.4.1 and Section 1.6]). Making use of these properties of virtual turning points, we obtain a recipe for finding the **Stokes geometry** for (1.1.1), that is, the figure that describes the location of relevant turning points (both ordinary and virtual) and Stokes curves emanating from them. See [VTP, Section 1.5] for the details.

Next, let us recall the definition of s-virtual turning points. The **s-virtual turning point** is a peculiar version of the virtual turning point when an integral representation of solutions is available. As the notion of the s-virtual turning point is defined based on the structure theorem for (micro)differential equations established in [SKK, Chapter II] and this theorem covers not only single equations but also over-determined systems, we explain the definition of s-virtual turning points in the framework of over-determined systems for our later convenience.

Fig. 1.1 Stokes geometry for the BNR equation

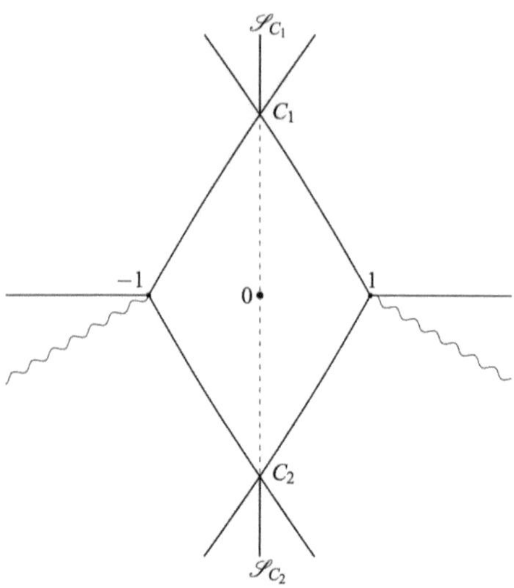

Let $\mathscr{M}$ be a holonomic system or maximally over-determined system with a large parameter η. A typical example is the so-called **Pearcey system** (cf. [A], [Hi2], etc.):

$$\begin{cases} \left(4\dfrac{\partial^3}{\partial x_1^3} + 2\eta^2 x_2 \dfrac{\partial}{\partial x_1} + \eta^3 x_1\right)\psi = 0 \\ \left(\dfrac{\partial^2}{\partial x_1^2} - \eta \dfrac{\partial}{\partial x_2}\right)\psi = 0, \end{cases} \tag{1.1.20}$$

which will be denoted by $\mathscr{M}_3$ in what follows since its particular solution given by the **Pearcey integral**

$$\psi(x, \eta) = \psi(x_1, x_2, \eta) = \int \exp \eta(t^4 + x_2 t^2 + x_1 t)\, dt \tag{1.1.21}$$

is related to the theory of singularities of type A_3. Note also that the restriction of (1.1.20) to $x_2 = c_2$ with non-zero constant c_2 is equivalent to the BNR equation (1.1.19). We now assume that a solution ψ of $\mathscr{M}$ enjoys the integral representation

$$\psi(x, \eta) = \int \exp(\eta F(x, t)) dt \tag{1.1.22}$$

for a polynomial $F(x, t)$ of $(x, t) \in \mathbb{C}^n \times \mathbb{C}^p$. Letting

$$\mathscr{C} = \left\{(x, t) \in \mathbb{C}^n \times \mathbb{C}^p\, ;\ \mathrm{grad}_t F(x, t) = 0, \text{ i.e.,} \frac{\partial F}{\partial t_1} = \cdots = \frac{\partial F}{\partial t_p} = 0\right\} \tag{1.1.23}$$

denote the set of critical points or **saddle points** of $F(x, t)$ with respect to t, we also assume that $F(x, t)$ satisfies the following condition:

$$\varpi : \mathscr{C} \longrightarrow \mathbb{C}^n_x \text{ is a finite proper map,} \tag{1.1.24}$$

that is, ϖ maps a closed set to a closed set, and for each x_0 in $\mathbb{C}^n_x$ there exist an open neighborhood ω of x_0 and continuous vector functions $t^{(j)}(x) = (t_1^{(j)}(x), \ldots, t_p^{(j)}(x))$ $(j = 1, 2, \ldots, N)$ on ω for which the following holds:

$$\varpi^{-1}(\omega) \subset \bigcup_{j=1}^{N} \{(x, t) \in \omega \times \mathbb{C}^p\, ;\ t = t^{(j)}(x)\}. \tag{1.1.25}$$

In this setting the Borel transform $\psi_B(x, y)$ of $\psi(x, \eta)$ is given by

$$\psi_B(x, y) = \int \exp(y\eta)\Big(\int \exp\big(\eta F(x, t)\big)dt\Big)d\eta = \int \delta\big(y + F(x, t)\big)dt, \tag{1.1.26}$$

where δ denotes the delta function. Note that $\phi(t, x, y) = \delta(y + F(x, t))$ satisfies a maximally over-determined system

$$(y + F(x, t))\phi = 0, \quad \frac{\partial}{\partial t_j}\phi - \frac{\partial F}{\partial t_j}\frac{\partial}{\partial y}\phi = 0 \quad (j = 1, \ldots, p),$$

$$\frac{\partial}{\partial x_k}\phi - \frac{\partial F}{\partial x_k}\frac{\partial}{\partial y}\phi = 0 \quad (k = 1, \ldots, n),$$

and its characteristic variety (outside the zero section) is contained in

$$\Big\{(t, x, y; \tau, \xi, \eta) \in T^*\mathbb{C}^{p+n+1}_{(t,x,y)}\ ;\ y = -F(x, t) \text{ and}$$
$$(\tau, \xi, \eta) = c(\mathrm{grad}_t F(x, t), \mathrm{grad}_x F(x, t), 1) \text{ for a non-zero complex number } c\Big\}.$$

Then the general theory of "**integration (along fibers) of (over-determined) systems**" developed in [SKK, Chapter II] tells us that the characteristic variety (outside the zero section) of the maximally over-determined system that $\psi_B(x, y)$ satisfies is contained in

$$V = \Big\{(x, y; \xi, \eta) \in T^*\mathbb{C}^{n+1}_{(x,y)}\ ;\ y = -F(x, t) \text{ with } \mathrm{grad}_t F(x, t) = 0 \text{ and}$$
$$(\xi, \eta) = c(\mathrm{grad}_x F(x, t), 1) \text{ for a non-zero complex number } c\Big\}. \tag{1.1.27}$$

In particular, the projection of V to the base manifold $\mathbb{C}^{n+1}_{(x,y)}$ is contained in

$$\mathscr{S} = \Big\{(x, y) \in \mathbb{C}^{n+1}\ ;\ \text{there exists } (x, t) \in \mathscr{C} \text{ for which } y = -F(x, t) \text{ holds}\Big\}. \tag{1.1.28}$$

This means that the singularities of $\psi_B(x, y)$ correspond to the saddle points $(x, t) \in \mathscr{C}$ through the mapping $y = F(x, t)$ for each $x \in \mathbb{C}^n$. As all points in V (and those in (1.1.28) as well) are thought to be "cognate", an s-virtual turning point is defined as follows:

Definition 1.5 A point x_0 is said to be an s**-virtual turning point** of $\mathscr{M}$ if there exist t and t' with $t \neq t'$ for which the following conditions are satisfied:

$$F(x_0, t) = F(x_0, t'), \tag{1.1.29}$$

$$(\mathrm{grad}_t F)(x_0, t) = (\mathrm{grad}_t F)(x_0, t') = 0. \tag{1.1.30}$$

That is, just like a virtual turning point, at an s-virtual turning point x_0 two cognate singularities contained in (1.1.28), i.e., the projection of V to the base manifold

$\mathbb{C}^{n+1}_{(x,y)}$ merge. Since points in $\mathscr{C}$ are labeled as $t = t^{(j)}(x)$ in a neighborhood of x_0 in view of (1.1.25), we can find two different labels j and k which satisfy

$$t^{(j)}(x_0) = t \text{ and } t^{(k)}(x_0) = t', \tag{1.1.31}$$

and hence a new Stokes curve (or, more appropriately, "**new Stokes surface**" in this case) emanating from an s-virtual turning point x_0 is given by

$$\text{Im } F(x, t^{(j)}(x)) = \text{Im } F(x, t^{(k)}(x)) \tag{1.1.32}$$

near x_0.

Remark 1.2 In this volume we use the notation $F(x, t)$ to denote a phase function of the integral representation (1.1.22) instead of $S(x, t)$ used in [VTP]. We also use $\exp(\eta y)$ in (1.1.26) instead of $\exp(-\eta y)$ used in [VTP] and consequently $S(x, t)$ in [VTP, (1.8.12)] is changed to $-F(x, t)$ in the above discussion. Although there is no essential difference between these two notations, we note this to avoid the possible confusion of the reader.

In the case of the Pearcey system (1.1.20) the Pearcey integral (1.1.21) gives an integral representation of solutions. Hence $F(x, t) = t^4 + x_2t^2 + x_1t$, $n = 2$ and $p = 1$. Let us compute the s-virtual turning points of the Pearcey system (1.1.20) by using (1.1.21). For (1.1.21) the condition (1.1.29) reads as

$$t^4 + x_2t^2 + x_1t = t'^4 + x_2t'^2 + x_1t', \tag{1.1.33}$$

which together with $t \neq t'$ implies

$$(t + t')(t^2 + t'^2) + x_2(t + t') + x_1 = 0. \tag{1.1.34}$$

On the other hand, (1.1.30) gives

$$4t^3 + 2x_2t + x_1 = 0 \text{ and } 4t'^3 + 2x_2t' + x_1 = 0. \tag{1.1.35}$$

Summing up these two equations, we find

$$2(t + t')(t^2 - tt' + t'^2) + x_2(t + t') + x_1 = 0. \tag{1.1.36}$$

Then it follows from (1.1.34) and (1.1.36) that

$$(t + t')(t^2 - 2tt' + t'^2) = (t + t')(t - t')^2 = 0, \tag{1.1.37}$$

which leads to $t + t' = 0$. Hence by (1.1.34) we obtain $x_1 = 0$, and also $x_2 \neq 0$ by (1.1.35). Thus a point (x_1, x_2) with $x_1 = 0$ (and $x_2 \neq 0$) is an s-virtual turning point of (1.1.20). As this computation shows, s-virtual turning points are computed by simple algebraic manipulations, that is, more easily than virtual turning points when an integral representation of solutions is available.

It is expected that the union of all the ordinary turning points and all the s-virtual turning points coincides with the totality of ordinary and virtual turning points. Since in this volume we mainly discuss higher-order ordinary differential equations or rather over-determined systems and especially their tangential systems when an integral representation of solutions is available, in what follows s-virtual turning points play a more important role than virtual turning points in some respects.

1.2 Appearance of a Non-hereditary Turning Point Through a Tangential System

One of the main purposes of this volume is to study virtual turning points of a system of differential equations. In particular, we want to consider the following problem:

How are the (ordinary and virtual) turning points of a system of differential equations related with those of its tangential system?

Since the virtual turning points are defined through the microlocal analysis, that is, in terms of the bicharacteristic strips of a system, it is expected that they should be well behaved under such an operation as taking a tangential system. However, the situation is not so simple. In this chapter we discuss this problem by taking the Pearcey system

$$\mathcal{M}_3: \quad \begin{cases} \left(4\eta^{-3}\dfrac{\partial^3}{\partial x_1^3} + 2\eta^{-1}x_2\dfrac{\partial}{\partial x_1} + x_1\right)\psi = 0 \\ \left(\eta^{-1}\dfrac{\partial}{\partial x_2} - \eta^{-2}\dfrac{\partial^2}{\partial x_1^2}\right)\psi = 0 \end{cases} \tag{1.2.1}$$

as a typical example.

Defining an unknown vector Ψ by $\Psi = {}^t(\psi, \eta^{-1}(\partial\psi/\partial x_1), \eta^{-1}(\partial\psi/\partial x_2)) = {}^t(\psi, \eta^{-1}(\partial\psi/\partial x_1), \eta^{-2}(\partial^2\psi/\partial x_1^2))$ (where ^{t} designates the transposition), we find the Pearcey system can be written also in the following form of a completely integrable system of first-order equations:

$$\mathcal{M}_3: \quad \begin{cases} \eta^{-1}\dfrac{\partial}{\partial x_1}\Psi = P_0\Psi \\ \eta^{-1}\dfrac{\partial}{\partial x_2}\Psi = (Q_0 + \eta^{-1}Q_1)\Psi, \end{cases} \tag{1.2.2}$$

where

$$P_0 = \begin{pmatrix} 0 & 1 & 0 \\ 0 & 0 & 1 \\ -\frac{x_1}{4} & -\frac{x_2}{2} & 0 \end{pmatrix}, \tag{1.2.3}$$

$$Q_0 = P_0^2 = \begin{pmatrix} 0 & 0 & 1 \\ -\frac{x_1}{4} & -\frac{x_2}{2} & 0 \\ 0 & -\frac{x_1}{4} & -\frac{x_2}{2} \end{pmatrix}, \quad Q_1 = \frac{\partial P_0}{\partial x_1} = \begin{pmatrix} 0 & 0 & 0 \\ 0 & 0 & 0 \\ -\frac{1}{4} & 0 & 0 \end{pmatrix}. \tag{1.2.4}$$

(We designate the system (1.2.2) by the same symbol $\mathcal{M}_3$ as (1.2.1).) We now study the tangential system of the Pearcey system $\mathcal{M}_3$ restricted to the hyperplane

$$Y(c) = \{(x_1, x_2) \in \mathbb{C}^2\,;\ x_2 = c(x_1 - 1)\} \tag{1.2.5}$$

for a non-zero constant c.

In this volume we use the general notation $\mathcal{N}[f_j = 0]$ to denote the tangential system of a holonomic system $\mathcal{M}$ restricted to a submanifold defined by a set of equations $f_j = 0$. For example, the tangential system of $\mathcal{M}_3$ restricted to $Y(c)$ is denoted by $\mathcal{N}_3[x_2 = c(x_1 - 1)]$. For the sake of simplification of notation we abbreviate $\mathcal{N}_3[x_2 = c(x_1 - 1)]$ as $\mathcal{N}(c)$ in what follows.

If we introduce new variables (z_1, z_2) defined by

$$z_1 = x_1, \qquad z_2 = x_2 - c(x_1 - 1), \tag{1.2.6}$$

then $\mathcal{M}_3$ is transformed to

$$\widetilde{\mathcal{M}}_3 : \quad \begin{cases} \eta^{-1}\dfrac{\partial}{\partial z_1}\Psi = ((P_0 + cQ_0) + \eta^{-1}cQ_1)\Psi \\[2mm] \eta^{-1}\dfrac{\partial}{\partial z_2}\Psi = (Q_0 + \eta^{-1}Q_1)\Psi. \end{cases} \tag{1.2.7}$$

Hence its tangential system restricted to $Y(c)$ is expressed as

$$\mathcal{N}(c) : \quad \eta^{-1}\frac{d}{dz_1}\Psi = ((P_0 + cQ_0) + \eta^{-1}cQ_1)\Psi. \tag{1.2.8}$$

Here we denote the restriction of the matrices P_0, Q_0 and Q_1 to $Y(c)$ by the same symbols to avoid complicated notation. We hope there is no fear of confusion to the reader.

We first examine the turning points of the Pearcey system $\mathcal{M}_3$ and those of its tangential system $\mathcal{N}(c)$. As is discussed by Aoki [A] and one of the authors [Hi1], a turning point of $\mathcal{M}_3$ is, by definition, a common zero of the discriminants of two

characteristic equations $\det(\zeta - P_0) = 0$ and $\det(\zeta - Q_0) = 0$, where P_0 and Q_0 are the top-order coefficient matrices of (1.2.2) with respect to η. Since $Q_0 = P_0^2$ holds in this case, a turning point of $\mathscr{M}_3$ is consequently a zero of the discriminant of

$$\det(\zeta - P_0) = \zeta^3 + \frac{x_2}{2}\zeta + \frac{x_1}{4} = 0. \tag{1.2.9}$$

As the discriminant of (1.2.9) is explicitly given by $-(27x_1^2 + 8x_2^3)/16$, we thus obtain

$$\{\text{turning point of } \mathscr{M}_3\} = \{(x_1, x_2) \in \mathbb{C}^2 \,;\, 27x_1^2 + 8x_2^3 = 0\}. \tag{1.2.10}$$

Let us compute turning points of $\mathscr{N}(c)$. A turning point of $\mathscr{N}(c)$ is a zero of the discriminant of $\det(\zeta - (P_0 + cQ_0)) = 0$, i.e., the characteristic equation of the top-order coefficient matrix of (1.2.8). If we let ζ_j $(j = 1, 2, 3)$ denote **characteristic roots** (i.e., eigenvalues) of P_0, then characteristic roots of $P_0 + cQ_0$ are given by $\zeta_j + c\zeta_j^2$ in view of $Q_0 = P_0^2$. Hence it follows from the definition of the discriminant that

$$\begin{aligned}
&(\text{discriminant of } \det(\zeta - (P_0 + cQ_0)) = 0) \\
&\quad = \prod_{j>k} \left((\zeta_j + c\zeta_j^2) - (\zeta_k + c\zeta_k^2)\right)^2 \\
&\quad = \prod_{j>k} \left((\zeta_j - \zeta_k)(1 + c(\zeta_j + \zeta_k))\right)^2 \\
&\quad = \prod_{j>k} (\zeta_j - \zeta_k)^2 \prod_{j>k} (1 + c(\zeta_j + \zeta_k))^2.
\end{aligned} \tag{1.2.11}$$

The first factor exactly coincides with the discriminant of $\det(\zeta - P_0) = 0$, that is, $-(27x_1^2 + 8x_2^3)/16$. On the other hand, since $\zeta_1 + \zeta_2 + \zeta_3 = 0$ holds for the characteristic roots of P_0, i.e., the roots of (1.2.9), (the square-root of) the second factor can be computed as

$$\begin{aligned}
\prod_{j>k} (1 + c(\zeta_j + \zeta_k)) &= (1 - c\zeta_1)(1 - c\zeta_2)(1 - c\zeta_3) \\
&= c^3(c^{-1} - \zeta_1)(c^{-1} - \zeta_2)(c^{-1} - \zeta_3) \\
&= c^3\left(c^{-3} + \frac{x_2}{2}c^{-1} + \frac{x_1}{4}\right) \\
&= 1 + \frac{x_2}{2}c^2 + \frac{x_1}{4}c^3.
\end{aligned} \tag{1.2.12}$$

Hence we have

$$
\begin{aligned}
&(\text{discriminant of } \det(\zeta - (P_0 + cQ_0)) = 0) \\
&\quad = -\frac{1}{16}(27x_1^2 + 8x_2^3)\big(1 + \frac{x_2}{2}c^2 + \frac{x_1}{4}c^3\big)^2.
\end{aligned} \tag{1.2.13}
$$

Thus, in addition to the turning points of the Pearcey system $\mathcal{M}_3$, the tangential system $\mathcal{N}(c)$ also has a turning point at a zero of the second factor

$$
\begin{aligned}
1 + \frac{x_2}{2}c^2 + \frac{x_1}{4}c^3\bigg|_{z_2=0} &= 1 + \frac{z_2}{2}c^2 + \big(\frac{3}{4}z_1 - \frac{1}{2}\big)c^3\bigg|_{z_2=0} \\
&= c^3\left(\frac{3}{4}z_1 - \frac{1}{2} + \frac{1}{c^3}\right) = 0,
\end{aligned} \tag{1.2.14}
$$

that is, at $z_1 = (2 - 4c^{-3})/3$, which will be denoted by $\hat{a}_1$ in what follows. We call this additional turning point $z_1 = \hat{a}_1$ of $\mathcal{N}(c)$ a **non-hereditary turning point** (sometimes abbreviated as **NHTP** in what follows). A non-hereditary turning point is a turning point which is not observed for the original Pearcey system $\mathcal{M}_3$. It is a **double turning point** of $\mathcal{N}(c)$ in view of (1.2.13).

As is shown by the above computation for the Pearcey system $\mathcal{M}_3$, a non-hereditary turning point appears when we consider the tangential system. In the case of $\mathcal{M}_3$ and its tangential system $\mathcal{N}(c)$, the following assertions (a) and (b) hold at the NHTP $z_1 = \hat{a}_1$:

(a) the characteristic roots ζ_j $(j = 1, 2, 3)$ of P_0 are mutually distinct,
(b) $1 + c(\zeta_j + \zeta_k) = 0$ for some $j \neq k$ $(j, k \in \{1, 2, 3\})$.

The assertion (b) implies

(b)′ $\zeta_j + c\zeta_j^2 = \zeta_k + c\zeta_k^2$ for some $j \neq k$ $(j, k \in \{1, 2, 3\})$.

Otherwise stated, at the NHTP $z_1 = \hat{a}_1$ two different characteristic roots of the first Eq. (1.2.7) merge while all characteristic roots of the second Eq. (1.2.7) remain distinct (cf. Fig. 1.2). Two characteristic roots of only one of the equations of a system merge at a NHTP.

The above observation explains the mechanism for the appearance of a non-hereditary turning point. Having this mechanism in mind, we expect that a non-hereditary turning point appears in general when we consider a tangential system and plays an important role for its analysis. However, a non-hereditary turning point is irrelevant to the Stokes phenomena of WKB solutions of a tangential system near the point. This peculiar property of non-hereditary turning points is verified by using a result of [T3] on the diagonalization of completely integrable systems. Here let us confirm this fact for the Pearcey system $\mathcal{M}_3$ and its tangential system $\mathcal{N}(c)$.

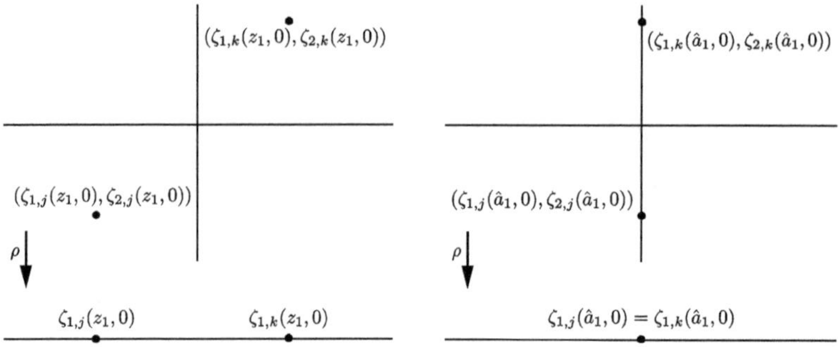

Fig. 1.2 Schematic correspondence between characteristic roots of $\mathscr{M}_3$ and those of $\mathscr{N}(c)$ near a NHTP $z_1 = \hat{a}_1$. Left-hand (resp., right-hand) figure designates the configuration at $z_1 \neq \hat{a}_1$ (resp., $z_1 = \hat{a}_1$). Here $\zeta_{1,j}(z_1, z_2)$ (resp., $\zeta_{2,j}(z_1, z_2)$) denotes a characteristic root of the first (resp., second) equation of (1.2.7)

Let

$$(a_1(z_2, c), z_2) = \Big(\frac{1}{3}\big(2 - \frac{4}{c^3} - \frac{2z_2}{c}\big), z_2\Big) \tag{1.2.15}$$

be a zero of the second factor of (1.2.13) so that $\hat{a}_1 = a_1(0, c)$ becomes the NHTP of $\mathscr{N}(c)$. Arguments similar to the preceding paragraph tell us that the characteristic roots ζ_j of P_0 and ζ_j^2 of Q_0 $(j = 1, 2, 3)$ are mutually distinct, respectively, while two characteristic roots $\zeta_j + c\zeta_j^2$ and $\zeta_k + c\zeta_k^2$ $(j, k \in \{1, 2, 3\}, j \neq k)$ of $P_0 + cQ_0$ merge at $z = (z_1, z_2) = (a_1(z_2, c), z_2)$. We assume $\zeta_1 + c\zeta_1^2$ and $\zeta_2 + c\zeta_2^2$ merge and $\zeta_3 + c\zeta_3^2$ remain distinct at $z = (a_1(z_2, c), z_2)$ in what follows. Then, by a result on the block-diagonalization of completely integrable systems (cf., e.g., [T3, Proposition 1]), near $z = (a_1(z_2, c), z_2)$ we can reduce the system $\widetilde{\mathscr{M}_3}$ by an appropriate gauge transformation of unknown functions

$$\tilde{\Psi} = T(z, \eta)\Psi, \quad T(z, \eta) = \sum_{n=0}^{\infty} \eta^{-n} T_n(z), \tag{1.2.16}$$

where $T_n(z)$ $(n = 0, 1, 2, \dots)$ are 3×3 matrices with holomorphic entries near $z = (a_1(z_2, c), z_2)$ satisfying $\det T_0(z) \neq 0$, to the following block-diagonalized system:

$$\begin{cases} \eta^{-1}\dfrac{\partial}{\partial z_1}\tilde{\Psi} = \begin{pmatrix} \tilde{R}(z,\eta) & 0 \\ 0 & \tilde{r}(z,\eta) \end{pmatrix} \tilde{\Psi} \\[2ex] \eta^{-1}\dfrac{\partial}{\partial z_2}\tilde{\Psi} = \begin{pmatrix} \tilde{S}(z,\eta) & 0 \\ 0 & \tilde{s}(z,\eta) \end{pmatrix} \tilde{\Psi}. \end{cases} \tag{1.2.17}$$

Here $\tilde{R}(z,\eta)$ and $\tilde{S}(z,\eta)$ are 2×2 matrices whose entries are power series in η^{-1},

$$\tilde{R}(z,\eta)=\sum_{n=0}^{\infty}\eta^{-n}\tilde{R}_n(z),\quad \tilde{S}(z,\eta)=\sum_{n=0}^{\infty}\eta^{-n}\tilde{S}_n(z),\tag{1.2.18}$$

and $\tilde{r}(z,\eta)$ and $\tilde{s}(z,\eta)$ are scalar power series in η^{-1},

$$\tilde{r}(z,\eta)=\sum_{n=0}^{\infty}\eta^{-n}\tilde{r}_n(z),\quad \tilde{s}(z,\eta)=\sum_{n=0}^{\infty}\eta^{-n}\tilde{s}_n(z).\tag{1.2.19}$$

Furthermore, they satisfy the following properties:

(i) The eigenvalues of $\tilde{R}_0(z)$ are $\zeta_1+c\zeta_1^2$ and $\zeta_2+c\zeta_2^2$. They merge at $z=(a_1(z_2,c),z_2)$.
(ii) $\tilde{r}_0(z)$ is distinct from the eigenvalues $\zeta_j+c\zeta_j^2$ $(j=1,2)$ of $\tilde{R}_0(z)$ near $z=(a_1(z_2,c),z_2)$.
(iii) The eigenvalues of $\tilde{S}_0(z)$ and $\tilde{s}_0(z)$ are mutually distinct near $z=(a_1(z_2,c),z_2)$.
(iv) $z=(a_1(z_2,c),z_2)$ is a double turning point of the first equation of (1.2.17) satisfying

$$\operatorname{rank}\left[\left(\tilde{R}_0(z)-\begin{pmatrix}\zeta_1+c\zeta_1^2 & 0\\ 0 & \zeta_2+c\zeta_2^2\end{pmatrix}\right)\Bigg|_{z=(a_1(z_2,c),z_2)}\right]=0.\tag{1.2.20}$$

Note that (1.2.20) follows from the compatibility condition of (1.2.17), i.e.,

$$\frac{\partial\tilde{R}}{\partial z_2}-\frac{\partial\tilde{S}}{\partial z_1}+\eta(\tilde{R}\tilde{S}-\tilde{S}\tilde{R})=0\ \text{ and }\ \frac{\partial\tilde{r}}{\partial z_2}-\frac{\partial\tilde{s}}{\partial z_1}=0,\tag{1.2.21}$$

in particular, its top-degree part $\tilde{R}_0\tilde{S}_0-\tilde{S}_0\tilde{R}_0=0$, and the property (iii). For more details see [Hi2, Theorem 4.1].

This block-diagonalization reduces the analysis of $\widetilde{\mathscr{M}}_3$ near $z=(a_1(z_2,c),z_2)$ to that of the following 2×2 system:

$$\begin{cases}\eta^{-1}\dfrac{\partial}{\partial z_1}\Phi=\tilde{R}(z,\eta)\Phi\\[2mm] \eta^{-1}\dfrac{\partial}{\partial z_2}\Phi=\tilde{S}(z,\eta)\Phi.\end{cases}\tag{1.2.22}$$

Then, thanks to the above properties (i)–(iv), we can apply a result of [T3, Theorem 3] to obtain the following:

Theorem 1.1 *The completely integrable system* (1.2.22) *can be transformed to*

$$\begin{cases} \eta^{-1}\dfrac{\partial}{\partial \tilde{z}_1}\tilde{\Phi} = \begin{pmatrix} -\tilde{z}_1 & 0 \\ 0 & \tilde{z}_1 \end{pmatrix} \tilde{\Phi} \\ \eta^{-1}\dfrac{\partial}{\partial \tilde{z}_2}\tilde{\Phi} = \begin{pmatrix} -1 & 0 \\ 0 & 1 \end{pmatrix} \tilde{\Phi} \end{cases} \tag{1.2.23}$$

by a change of independent variables $(\tilde{z}_1, \tilde{z}_2) = (\tilde{z}_1(z), \tilde{z}_2(z))$ *and a gauge transformation of the form*

$$\tilde{\Phi} = \exp(\eta f(z)) \left(\sum_{n=0}^{\infty} \eta^{-n} \tilde{T}_n(z) \right) \Phi, \tag{1.2.24}$$

where $f(z)$ *is a holomorphic function near* $z = (a_1(z_2, c), z_2)$ *and* $\tilde{T}_n(z)$ $(n = 0, 1, 2, \ldots)$ *are* 2×2 *matrices with holomorphic entries near* $z = (a_1(z_2, c), z_2)$ *satisfying* $\det \tilde{T}_0(z) \neq 0$. *Furthermore, the gauge transformation* (1.2.24) *is Borel transformable or Gevrey-1, that is, there exist positive constants A and C for which*

$$\|\tilde{T}_n(z)\| \leq AC^n n! \quad (n = 0, 1, 2, \ldots) \tag{1.2.25}$$

holds near $z = (a_1(z_2, c), z_2)$.

The canonical Eq. (1.2.23) is a direct sum of two trivial systems and free from any Stokes phenomena. Hence no Stokes phenomena are observed also for the first equation of $\widetilde{\mathcal{M}_3}$, i.e., the tangential system $\mathcal{N}(c)$. Thus the non-hereditary turning point $\hat{a}_1 = a_1(0, c)$ is irrelevant to the Stokes phenomena of WKB solutions of $\mathcal{N}(c)$. Otherwise stated, the non-hereditary turning point of $\mathcal{N}(c)$ is redundant and the Stokes curves emanating from it are inert.

Note that, in contrast to the redundancy of the NHTP $\hat{a}_1 = a_1(0, c)$ of $\mathcal{N}(c)$, it becomes non-redundant if we consider a perturbed system of $\mathcal{N}(c)$. This intriguing fact was suggested by the late Professor T. Koike and will be discussed in Chap. 3 of this volume.

1.3 A Non-hereditary Turning Point from the Viewpoint of an s-Virtual Turning Point

The Pearcey system $\mathcal{M}_3$ admits an integral representation of solutions through the Pearcey integral (1.1.21). As its consequence, the tangential system $\mathcal{N}(c)$ also has an integral representation

$$\psi(z_1, \eta) = \int \exp \eta(t^4 + c(z_1 - 1)t^2 + z_1 t)\, dt. \tag{1.3.1}$$

In this section we discuss the non-hereditary turning point of $\mathcal{N}(c)$ from the viewpoint of an s-virtual turning point and integral representation.

Let us first compute an s-virtual turning point of the tangential system $\mathcal{N}(c)$ in a way parallel to the computation of an s-virtual turning point of the Pearcey system $\mathcal{M}_3$ done in Sect. 1.1. In this section we abbreviate z_1 to z and let

$$F(z,t) = t^4 + c(z-1)t^2 + zt \tag{1.3.2}$$

be the phase function of (1.3.1). Then an s-virtual turning point of $\mathcal{N}(c)$ is determined by the equations

$$F(z,t) = F(z,t'), \text{ i.e., } t^4 + c(z-1)t^2 + zt = t'^4 + c(z-1)t'^2 + zt', \tag{1.3.3}$$

$$\frac{\partial F}{\partial t}(z,t) = 4t^3 + 2c(z-1)t + z = 0 \text{ and}$$
$$\frac{\partial F}{\partial t}(z,t') = 4t'^3 + 2c(z-1)t' + z = 0 \tag{1.3.4}$$

together $t \neq t'$. Parallel to (1.1.34), it follows from (1.3.3) that

$$(t+t')(t^2+t'^2) + c(z-1)(t+t') + z = 0, \tag{1.3.5}$$

while by summing up the two equations of (1.3.4) we find

$$2(t+t')(t^2 - tt' + t'^2) + c(z-1)(t+t') + z = 0. \tag{1.3.6}$$

Comparison of (1.3.5) with (1.3.6) implies

$$(t+t')(t^2 - 2tt' + t'^2) = (t+t')(t-t')^2 = 0. \tag{1.3.7}$$

Hence we obtain $t + t' = 0$ and $z = 0$ again. (Consequently we also obtain $c = 2t^2 = 2t'^2$ by (1.3.4).) Thus $z = 0$ is the unique s-virtual turning point of the tangential system $\mathcal{N}(c)$.

Next, we consider the non-hereditary turning point of $\mathcal{N}(c)$ by making full use of the integral representation (1.3.1). Since the tangential system $\mathcal{N}(c)$ has the integral representation (1.3.1), its characteristic variety and characteristic roots can be expressed in terms of it. For example, thanks to (1.1.27), the characteristic variety (outside the zero section) of the Borel transform of $\mathcal{N}(c)$ is contained in

$$\Big\{(z,y;\zeta,\eta) \in T^*\mathbb{C}^2\,;\, y = -F(z,t) \text{ with } \frac{\partial F}{\partial t}(z,t) = 0,$$
$$(\zeta,\eta) = \alpha\Big(\frac{\partial F}{\partial z}(z,t), 1\Big) \ (\alpha \neq 0)\Big\} \tag{1.3.8}$$

and its characteristic root is given by

$$\frac{\partial F}{\partial z}(z,t) \text{ with } \frac{\partial F}{\partial t}(z,t)=0. \tag{1.3.9}$$

We then find that z_* is an ordinary turning point if and only if there exist two branches $t_1(z)$ and $t_2(z)$ of the roots of $(\partial F/\partial t)(z,t_j(z))=0\ (j=1,2)$ satisfying

$$\frac{\partial F}{\partial z}(z_*,t_1(z_*))=\frac{\partial F}{\partial z}(z_*,t_2(z_*)). \tag{1.3.10}$$

From this viewpoint a non-hereditary turning point z_* is characterized as a point where

$$t_1(z_*)\neq t_2(z_*) \tag{1.3.11}$$

holds in addition to (1.3.10). In the case of $F(z,t)=t^4+c(z-1)t^2+zt$, $t_1(z)$ and $t_2(z)$ are two different roots of a cubic equation

$$\frac{\partial F}{\partial t}=4t^3+2c(z-1)t+z=0 \tag{1.3.12}$$

for t. Further, as $\partial F/\partial z=ct^2+t$, the condition (1.3.10) reads as

$$ct_1(z_*)^2+t_1(z_*)=ct_2(z_*)^2+t_2(z_*). \tag{1.3.13}$$

Hence, if z_* is a NHTP of $\mathcal{N}(c)$, we obtain from (1.3.11) and (1.3.13) that

$$c\big(t_1(z_*)+t_2(z_*)\big)+1=0. \tag{1.3.14}$$

Since the three roots $t_1(z)$, $t_2(z)$ and $t_3(z)$ of (1.3.12) satisfy $t_1(z)+t_2(z)+t_3(z)=0$, it follows from (1.3.14) that $-(t_1(z_*)+t_2(z_*))=c^{-1}$ gives the third root of (1.3.12) and hence satisfies (1.3.12) at $z=z_*$, that is,

$$4c^{-3}+2(z_*-1)+z_*=0. \tag{1.3.15}$$

Thus we obtain $z_*=(2-4c^{-3})/3$, the same result as in the preceding section.

As is clearly shown by this computation, a non-hereditary turning point can be located by using the integral representation of solutions. Comparing the above computation with that in Sect. 1.2, we find that the condition (1.3.10) or the equivalent conditions (1.3.13) and (1.3.14) correspond to the assertions (b)′ and (b) in Sect. 1.2, respectively, and further that the condition (1.3.11) corresponds to the assertion (a) in Sect. 1.2. When the integral representation of solutions is available, the computation of non-hereditary turning points can be done only by using the phase function of the integral representation.

1.4 Computation of the Characteristic Variety of $\mathscr{N}(c)$, the Tangential System of the Pearcey System $\mathscr{M}$ to the Hyperplane $Y(c)$

In this section we examine the characteristic variety and the bicharacteristic strip of the Borel transform of the tangential system $\mathscr{N}(c)$ to compare its virtual turning points and s-virtual turning points.

Since the tangential system $\mathscr{N}(c)$ is explicitly described by (1.2.8), its characteristic equation is given by

$$
\begin{aligned}
Q(z,\zeta,\eta) &= \det(\eta^{-1}\zeta - (P_0 + cQ_0))\Big|_{x_1=z,x_2=c(z-1)} \\
&= \det\begin{pmatrix} \eta^{-1}\zeta & -1 & -c \\ \dfrac{cz}{4} & \eta^{-1}\zeta + \dfrac{c^2(z-1)}{2} & -1 \\ \dfrac{z}{4} & \dfrac{c(z-1)}{2} + \dfrac{cz}{4} & \eta^{-1}\zeta + \dfrac{c^2(z-1)}{2} \end{pmatrix} \\
&= (\eta^{-1}\zeta)^3 + c^2(z-1)(\eta^{-1}\zeta)^2 + \frac{1}{4}(c^4(z-1)^2 + 5cz - 2c)(\eta^{-1}\zeta) \\
&\qquad + \frac{1}{16}(c^3z^2 - 2c^3z + 4z).
\end{aligned} \tag{1.4.1}
$$

Here the variable z_1 is abbreviated to z, ζ stands for the principal symbol of $\partial/\partial z$, and η is identified with the principal symbol of $\partial/\partial y$, where y denotes the variable dual to η through the Borel transform. Note that, in connection with the existence of the non-hereditary turning point, the characteristic equation $Q(z,\zeta,\eta) = 0$ of $\mathscr{N}(c)$ is not with simple characteristics. To avoid such a **double characteristics** property of the characteristic equation, we make use of the following factorization of $Q(z,\zeta,\eta)$: Completing the square with respect to z, we rewrite (1.4.1) as

$$
\begin{aligned}
Q(z,\zeta,\eta) = \frac{c^3}{16}\Bigg[&(4c\eta^{-1}\zeta + 1)\left(z + \frac{2c\eta^{-1}\zeta + 2 - c^3}{c^3}\right)^2 \\
&- \left(\frac{6c\eta^{-1}\zeta + 2 - c^3}{c^3}\right)^2\Bigg].
\end{aligned} \tag{1.4.2}
$$

In view of (1.4.2) we find $Q(z,\zeta,\eta)$ can be factorized as follows:

$$
Q(z,\zeta,\eta) = -\frac{1}{16c^3}q_+(z,\zeta,\eta)q_-(z,\zeta,\eta), \tag{1.4.3}
$$

where

$$q_{\pm}(z,\zeta,\eta)=(6c\eta^{-1}\zeta+2-c^3)\pm(2c\eta^{-1}\zeta+2-c^3+c^3z)\sqrt{4c\eta^{-1}\zeta+1}. \qquad (1.4.4)$$

The characteristic variety of the Borel transform of $\mathcal{N}(c)$ is then given by

$$-16c^3Q(z,\zeta,\eta)=q_+(z,\zeta,\eta)q_-(z,\zeta,\eta)=0. \qquad (1.4.5)$$

In particular, at a non-hereditary turning point $\hat{a}=(2-4c^{-3})/3$ computed in Sect. 1.2 (cf. (1.2.14)), the characteristic equation $Q=0$ has a double zero $\hat{\theta}=(c^2-2c^{-1})/6$. Although both q_+ and q_- vanish at $z=\hat{a}$ and $\eta^{-1}\zeta=\hat{\theta}$, it follows from the concrete expression (1.4.4) that each of $q_{\pm}$ is with simple characteristics. In what follows, to locate the virtual turning points of $\mathcal{N}(c)$, we compute the bicharacteristic strip of $q_{\pm}(z,\zeta,\eta)$ that emanates from $(z,\eta^{-1}\zeta)=(\hat{a},\hat{\theta})$ instead of the bicharacteristic strip of $Q(z,\zeta,\eta)$ itself.

Remark 1.3 The factorization (1.4.3)–(1.4.4) of the characteristic equation $Q(z,\zeta,\eta)=0$ is obtained also by using the characteristic equation

$$4(\eta^{-1}\xi_1)^3+2x_2(\eta^{-1}\xi_1)+x_1=0, \quad (\eta^{-1}\xi_2)-(\eta^{-1}\xi_1)^2=0 \qquad (1.4.6)$$

of the Pearcey system $\mathcal{M}_3$, where ξ_j $(j=1,2)$ stands for the principal symbol of $\partial/\partial x_j$ (cf. (1.2.1)). As a matter of fact, substituting $(x_1,x_2)=(z,c(z-1))$ and

$$\xi_1=\zeta_1-c\zeta_2, \quad \xi_2=\zeta_2, \qquad (1.4.7)$$

which corresponds to the change of variables (1.2.6), into (1.4.6) and eliminating ζ_2 from the resulting equations, we obtain the same factorization as (1.4.3)–(1.4.4) of the characteristic equation of $\mathcal{N}(c)$. For more details see [HiKT, Section 2.3].

Now let us compute the bicharacteristic strip of, say, $q_-(z,\zeta,\eta)$. (That of $q_+(z,\zeta,\eta)$ can be computed similarly.) To simplify the computation, we consider

$$q=\eta^{3/2}q_-=6c\eta^{1/2}\zeta+(2-c^3)\eta^{3/2}-(2c\zeta+(2-c^3+c^3z)\eta)\sqrt{4c\zeta+\eta} \qquad (1.4.8)$$

instead of q_-. Then the bicharacteristic strip of q is, by definition, a solution of the hamiltonian system

$$\begin{cases} \dfrac{dz}{d\sigma}=\dfrac{\partial q}{\partial\zeta}=6c\eta^{1/2}-2c\sqrt{4c\zeta+\eta}-(2c\zeta+(2-c^3+c^3z)\eta)\dfrac{2c}{\sqrt{4c\zeta+\eta}}, \\ \dfrac{dy}{d\sigma}=\dfrac{\partial q}{\partial\eta}=3c\eta^{-1/2}\zeta+\dfrac{3}{2}(2-c^3)\eta^{1/2}-(2-c^3+c^3z)\sqrt{4c\zeta+\eta} \\ \qquad\qquad -(2c\zeta+(2-c^3+c^3z)\eta)\dfrac{1}{2\sqrt{4c\zeta+\eta}}, \\ \dfrac{d\zeta}{d\sigma}=-\dfrac{\partial q}{\partial z}=c^3\eta\sqrt{4c\zeta+\eta}, \\ \dfrac{d\eta}{d\sigma}=-\dfrac{\partial q}{\partial y}=0 \end{cases} \qquad (1.4.9)$$

satisfying $q(z(\sigma), \zeta(\sigma), \eta(\sigma)) = 0$. Here, as an initial condition, we impose

$$(z(0), y(0), \zeta(0), \eta(0)) = (\hat{a}, 0, \hat{\theta}, 1) = \left(\frac{2 - 4c^{-3}}{3}, 0, \frac{c^2 - 2c^{-1}}{6}, 1\right). \tag{1.4.10}$$

Let us solve (1.4.9) under the initial condition (1.4.10). First, we find $\eta = 1$. Then the third equation of (1.4.9) implies

$$\frac{1}{\sqrt{4c\zeta + 1}} \frac{d\zeta}{d\sigma} = c^3. \tag{1.4.11}$$

This can be readily integrated as

$$\sqrt{4c\zeta + 1} = 2c^4(\sigma + \alpha) \tag{1.4.12}$$

with a constant α of integration, which is determined to be

$$\alpha = \frac{1}{2c^4}\left(\frac{2c^3 - 1}{3}\right)^{1/2} \tag{1.4.13}$$

by the initial condition (1.4.10). Letting $\tilde{\sigma}$ denote

$$\tilde{\sigma} = \sigma + \alpha, \tag{1.4.14}$$

we thus obtain

$$\sqrt{4c\zeta + 1} = 2c^4\tilde{\sigma}, \quad \zeta = c^7\tilde{\sigma}^2 - \frac{1}{4c}. \tag{1.4.15}$$

Next, we seek an explicit form of z. Substituting (1.4.15) into the first Eq. (1.4.9), we find

$$\frac{dz}{d\sigma} = \frac{dz}{d\tilde{\sigma}} = 6c - 6c^5\tilde{\sigma} + \left(1 - \frac{3}{2c^3}\right)\frac{1}{\tilde{\sigma}} - \frac{z}{\tilde{\sigma}}. \tag{1.4.16}$$

This can be rewritten as

$$\frac{d}{d\tilde{\sigma}}(\tilde{\sigma}z) = -6c^5\tilde{\sigma}^2 + 6c\tilde{\sigma} + \left(1 - \frac{3}{2c^3}\right). \tag{1.4.17}$$

Hence we have

$$z = -2c^5\tilde{\sigma}^2 + 3c\tilde{\sigma} + \left(1 - \frac{3}{2c^3}\right) + \frac{\beta}{\tilde{\sigma}} \tag{1.4.18}$$

with another constant β of integration. Since the initial condition (1.4.10) together with (1.4.13) and (1.4.14) implies

$$\beta = -3c\alpha^2 = \frac{1-2c^3}{4c^7}, \tag{1.4.19}$$

we obtain

$$z = -2c^5\tilde{\sigma}^2 + 3c\tilde{\sigma} + \left(1 - \frac{3}{2c^3}\right) + \left(-\frac{1}{2c^4} + \frac{1}{4c^7}\right)\frac{1}{\tilde{\sigma}}. \tag{1.4.20}$$

Finally, y is determined by

$$\frac{dy}{d\sigma} = \frac{dy}{d\tilde{\sigma}} = 4c^{12}\tilde{\sigma}^3 - 3c^8\tilde{\sigma}^2 - c^4\tilde{\sigma} + \left(1 - \frac{c^3}{2}\right) + \left(\frac{1}{8c^5} - \frac{1}{16c^8}\right)\frac{1}{\tilde{\sigma}^2}, \tag{1.4.21}$$

which follows from substitution of (1.4.15) and (1.4.20) into the second equation of (1.4.9). Taking the initial condition (1.4.10) into account, we thus obtain

$$y = c^{12}\tilde{\sigma}^4 - c^8\tilde{\sigma}^3 - \frac{1}{2}c^4\tilde{\sigma}^2 + \left(1 - \frac{c^3}{2}\right)\tilde{\sigma} + \left(-\frac{1}{8c^5} + \frac{1}{16c^8}\right)\frac{1}{\tilde{\sigma}} + \gamma \tag{1.4.22}$$

with

$$\gamma = (2c^3 - 1)\left(\frac{-2c^3 + 7}{144c^4} + \frac{\sqrt{3(2c^3 - 1)}}{18c^4}\right). \tag{1.4.23}$$

Summing up, (1.4.15), (1.4.20) and (1.4.22) describe the bicharacteristic strip of $q = q(z, \zeta, \eta)$. In particular,

$$\begin{cases} z(\tilde{\sigma}) = -2c^5\tilde{\sigma}^2 + 3c\tilde{\sigma} + \left(1 - \frac{3}{2c^3}\right) + \left(-\frac{1}{2c^4} + \frac{1}{4c^7}\right)\frac{1}{\tilde{\sigma}}, \\ y(\tilde{\sigma}) = c^{12}\tilde{\sigma}^4 - c^8\tilde{\sigma}^3 - \frac{1}{2}c^4\tilde{\sigma}^2 + \left(1 - \frac{c^3}{2}\right)\tilde{\sigma} + \left(-\frac{1}{8c^5} + \frac{1}{16c^8}\right)\frac{1}{\tilde{\sigma}} + \gamma \end{cases} \tag{1.4.24}$$

gives the bicharacteristic curve of q.

We now seek for a self-intersection point of the bicharacteristic curve (1.4.24) of q by solving

$$z(\sigma_1) = z(\sigma_2) \text{ and } y(\sigma_1) = y(\sigma_2) \text{ with } \sigma_1 \neq \sigma_2. \tag{1.4.25}$$

First, the equation $z(\sigma_1) = z(\sigma_2)$ implies

$$-2c^5\sigma_1^2 + 3c\sigma_1 + \left(-\frac{1}{2c^4} + \frac{1}{4c^7}\right)\frac{1}{\sigma_1} = -2c^5\sigma_2^2 + 3c\sigma_2 + \left(-\frac{1}{2c^4} + \frac{1}{4c^7}\right)\frac{1}{\sigma_2}, \tag{1.4.26}$$

that is,

$$-2c^4(\sigma_1+\sigma_2)+3+\Big(\frac{1}{2c^5}-\frac{1}{4c^8}\Big)\frac{1}{\sigma_1\sigma_2}=0. \tag{1.4.27}$$

Letting s and p denote $s=\sigma_1+\sigma_2$ and $p=\sigma_1\sigma_2$, respectively, we have

$$\frac{1}{4c^8}(2c^3-1)\frac{1}{p}=2c^4s-3. \tag{1.4.28}$$

Next, $y(\sigma_1)=y(\sigma_2)$ implies

$$\begin{aligned}c^{12}\sigma_1^4-c^8\sigma_1^3-\frac{1}{2}c^4\sigma_1^2+\Big(1-\frac{c^3}{2}\Big)\sigma_1+\Big(-\frac{1}{8c^5}+\frac{1}{16c^8}\Big)\frac{1}{\sigma_1}\\ =c^{12}\sigma_2^4-c^8\sigma_2^3-\frac{1}{2}c^4\sigma_2^2+\Big(1-\frac{c^3}{2}\Big)\sigma_2+\Big(-\frac{1}{8c^5}+\frac{1}{16c^8}\Big)\frac{1}{\sigma_2},\end{aligned} \tag{1.4.29}$$

that is,

$$\begin{aligned}c^{12}(\sigma_1+\sigma_2)(\sigma_1^2+\sigma_2^2)-c^8(\sigma_1^2+\sigma_1\sigma_2+\sigma_2^2)-\frac{1}{2}c^4(\sigma_1+\sigma_2)\\ +\Big(1-\frac{c^3}{2}\Big)+\Big(\frac{1}{8c^5}-\frac{1}{16c^8}\Big)\frac{1}{\sigma_1\sigma_2}=0.\end{aligned} \tag{1.4.30}$$

Substitution of (1.4.28) into (1.4.30) deduces

$$c^{12}s(s^2-2p)-c^8(s^2-p)-\frac{1}{2}c^4s+\Big(1-\frac{c^3}{2}\Big)+\frac{1}{4}(2c^4s-3)=0. \tag{1.4.31}$$

Hence we obtain

$$c^{12}s^3-c^8s^2+\frac{1}{4}-\frac{c^3}{2}=c^8p(2c^4s-1). \tag{1.4.32}$$

Combining (1.4.28) and (1.4.32), we find

$$4c^8s^2(c^4s-1)-(2c^3-1)=\frac{(2c^3-1)(2c^4s-1)}{2c^4s-3}, \tag{1.4.33}$$

which can be rewritten as

$$(c^4s-1)\big[c^8s^2(2c^4s-3)-(2c^3-1)\big]=0. \tag{1.4.34}$$

If we assume the second factor of (1.4.34) vanishes, then it follows from (1.4.28) that $s^2 = 4p$, which means $\sigma_1 = \sigma_2$. Thus, to seek for a self-intersection point of the bicharacteristic curve of q, we need to choose $c^4 s - 1 = 0$, that is,

$$s = \frac{1}{c^4} \text{ and } p = \frac{1-2c^3}{4c^8} \tag{1.4.35}$$

due to (1.4.28). Hence σ_1 and σ_2 are the roots of

$$4c^8\sigma^2 - 4c^4\sigma + (1-2c^3) = 0, \tag{1.4.36}$$

that is,

$$\sigma_1, \sigma_2 = \frac{1 \pm \sqrt{2c^3}}{2c^4}, \tag{1.4.37}$$

and $z(\sigma_1)$ (equivalently, $z(\sigma_2)$) is given by

$$\begin{aligned} z(\sigma_1) &= -2c^5\sigma_1^2 + 3c\sigma_1 + \Big(1 - \frac{3}{2c^3}\Big) + \Big(-\frac{1}{2c^4} + \frac{1}{4c^7}\Big)\frac{1}{\sigma_1} \\ &= \frac{1}{4c^7\sigma_1}\big(4c^8\sigma_1^2 - 4c^4\sigma_1 + (1-2c^3)\big)(-2c^4\sigma_1 + 1) \\ &= 0. \end{aligned} \tag{1.4.38}$$

This confirms the fact that $z = 0$ is a virtual turning point of $\mathcal{N}(c)$. The s-virtual turning point detected in Sect. 1.3 is also a virtual turning point in the sense of Definition 1.4. Compared to the s-virtual turning point, computation of the virtual turning point is very hard, as is shown by the above discussion.

In addition to the computation of virtual turning points of $\mathcal{N}(c)$, let us also compute virtual turning points of the Pearcey system $\mathcal{M}_3$ here. For that purpose we need to consider the **bicharacteristic manifold** of $\mathcal{M}_3$. Since $\mathcal{M}_3$ is given by (1.2.1), its bicharacteristic manifold is defined by the following hamiltonian system in the cotangent bundle $T^*\mathbb{C}^3_{(x_1,x_2,y)} = \mathbb{C}^6_{(x_1,x_2,y;\xi_1,\xi_2,\eta)}$:

$$\begin{aligned} &\frac{dx_1}{ds} = \frac{\partial H_1}{\partial \xi_1} = 12\xi_1^2 + 2x_2\eta^2, \quad \frac{dx_2}{ds} = \frac{\partial H_1}{\partial \xi_2} = 0, \\ &\frac{dy}{ds} = \frac{\partial H_1}{\partial \eta} = 4x_2\xi_1\eta + 3x_1\eta^2, \\ &\frac{d\xi_1}{ds} = -\frac{\partial H_1}{\partial x_1} = -\eta^3, \quad \frac{d\xi_2}{ds} = -\frac{\partial H_1}{\partial x_2} = -2\xi_1\eta^2, \quad \frac{d\eta}{ds} = -\frac{\partial H_1}{\partial y} = 0 \end{aligned} \tag{1.4.39}$$

and

$$\frac{dx_1}{dt} = \frac{\partial H_2}{\partial \xi_1} = -2\xi_1, \quad \frac{dx_2}{dt} = \frac{\partial H_2}{\partial \xi_2} = \eta, \quad \frac{dy}{dt} = \frac{\partial H_2}{\partial \eta} = \xi_2,$$

$$\frac{d\xi_1}{dt} = -\frac{\partial H_2}{\partial x_1} = 0, \quad \frac{d\xi_2}{dt} = -\frac{\partial H_2}{\partial x_2} = 0, \quad \frac{d\eta}{dt} = -\frac{\partial H_2}{\partial y} = 0, \tag{1.4.40}$$

where the hamiltonians H_1 and H_2 are given by

$$H_1 = 4\xi_1^3 + 2x_2\xi_1\eta^2 + x_1\eta^3, \quad H_2 = \xi_2\eta - \xi_1^2, \tag{1.4.41}$$

respectively. The initial condition is imposed at a simple turning point as follows:

$$(x_1, x_2, y, \xi_1, \xi_2, \eta)\Big|_{s=t=0} = \left(1, -\frac{3}{2}, 0, \frac{1}{2}, \frac{1}{4}, 1\right). \tag{1.4.42}$$

Then a solution of (1.4.40) with the initial condition (1.4.42) is given by

$$\xi_1 = \frac{1}{2}, \quad \xi_2 = \frac{1}{4}, \quad \eta = 1,$$

$$x_1 = -t + 1, \quad x_2 = t - \frac{3}{2}, \quad y = \frac{1}{4}t. \tag{1.4.43}$$

We next solve (1.4.39), regarding (1.4.43) as an initial condition. The result is

$$\xi_1 = -\left(s - \frac{1}{2}\right), \quad \xi_2 = \left(s - \frac{1}{2}\right)^2, \quad \eta = 1,$$

$$x_1 = 4\left(s - \frac{1}{2}\right)^3 + 2\left(t - \frac{3}{2}\right)\left(s - \frac{1}{2}\right), \quad x_2 = t - \frac{3}{2}, \tag{1.4.44}$$

$$y = 3\left(s - \frac{1}{2}\right)^4 + \left(t - \frac{3}{2}\right)\left(s - \frac{1}{2}\right)^2 + \frac{3}{16}.$$

Hence the projection of the bicharacteristic manifold to the base space $\mathbb{C}^3_{(x_1,x_2,y)}$ is expressed as

$$\begin{cases} x_1(\tilde{s}, \tilde{t}) = 4\tilde{s}^3 + 2\tilde{s}\tilde{t}, \\ x_2(\tilde{s}, \tilde{t}) = \tilde{t}, \\ y(\tilde{s}, \tilde{t}) = 3\tilde{s}^4 + \tilde{s}^2\tilde{t} + \dfrac{3}{16} \end{cases} \tag{1.4.45}$$

with $\tilde{s} = s - 1/2$ and $\tilde{t} = t - 3/2$.

Remark 1.4 The bicharacteristic manifold of $\mathscr{M}_3$ can be computed also by using the expression (1.2.2) of $\mathscr{M}_3$ in the form of a completely integrable system. In this case we need to solve the hamiltonian system

$$\frac{dx_j}{d\sigma} = \frac{\partial K_1}{\partial \xi_j}, \quad \frac{dy}{d\sigma} = \frac{\partial K_1}{\partial \eta}, \quad \frac{d\xi_j}{d\sigma} = -\frac{\partial K_1}{\partial x_j}, \quad \frac{d\eta}{d\sigma} = -\frac{\partial K_1}{\partial y},$$
$$\frac{dx_j}{d\tau} = \frac{\partial K_2}{\partial \xi_j}, \quad \frac{dy}{d\tau} = \frac{\partial K_2}{\partial \eta}, \quad \frac{d\xi_j}{d\tau} = -\frac{\partial K_2}{\partial x_j}, \quad \frac{d\eta}{d\tau} = -\frac{\partial K_2}{\partial y}, \tag{1.4.46}$$

($j = 1, 2$) with the hamiltonians

$$K_1 = \xi_1^3 + \frac{x_2}{2}\xi_1\eta^2 + \frac{x_1}{4}\eta^3, \quad K_2 = \xi_2^3 + x_2\xi_2^2\eta + \frac{x_2^2}{4}\xi_2\eta^2 - \frac{x_1^2}{16}\eta^3. \tag{1.4.47}$$

By straightforward computations we find that a solution of (1.4.46) with the initial condition (1.4.42) is given by

$$\xi_1 = -\frac{1}{4}\sigma + \frac{1}{8}\tilde{\tau}, \quad \xi_2 = \frac{1}{16}\Big(\sigma - \frac{1}{2}\tilde{\tau}\Big)^2, \quad \eta = 1,$$
$$x_1 = \frac{1}{16}\sigma^3 - \frac{3}{32}\sigma^2\tilde{\tau} + \frac{1}{32}\sigma\tilde{\tau}^2 - 2\frac{\sigma}{\tilde{\tau}} + 1, \quad x_2 = -\frac{4}{\tilde{\tau}} - \frac{1}{32}\tilde{\tau}^2, \tag{1.4.48}$$
$$y = \frac{3}{256}\sigma^4 - \frac{3}{128}\sigma^3\tilde{\tau} + \sigma^2\Big(\frac{1}{64}\tilde{\tau}^2 - \frac{1}{4\tilde{\tau}}\Big) - \frac{\sigma}{4}\Big(\frac{1}{64}\tilde{\tau}^3 - 1\Big) + \frac{\tilde{\tau}^4}{4096} - \frac{\tilde{\tau}}{16} + \frac{3}{16},$$

where $\tilde{\tau} = \tau + 4$. Comparing (1.4.44) and (1.4.48), we can confirm that (1.4.44) and (1.4.48) exactly coincide through a relation

$$\tilde{s} = \frac{1}{4}\sigma - \frac{1}{8}\tilde{\tau}, \quad \tilde{t} = -\frac{1}{32}\tilde{\tau}^2 - \frac{4}{\tilde{\tau}}. \tag{1.4.49}$$

However, the computation of the bicharacteristic manifold based on (1.2.1) is much simpler than that based on (1.2.2).

Using the expression (1.4.45), let us seek for a self-intersection point of the bicharacteristic manifold of the Pearcey system $\mathscr{M}_3$. To this end, we solve the equation

$$x_j(\tilde{s}_1, \tilde{t}_1) = x_j(\tilde{s}_2, \tilde{t}_2) \ (j = 1, 2) \quad \text{and} \quad y(\tilde{s}_1, \tilde{t}_1) = y(\tilde{s}_2, \tilde{t}_2) \tag{1.4.50}$$

with $(\tilde{s}_1, \tilde{t}_1) \neq (\tilde{s}_2, \tilde{t}_2)$. First, it follows from $x_2(\tilde{s}_1, \tilde{t}_1) = x_2(\tilde{s}_2, \tilde{t}_2)$ that $\tilde{t}_1 = \tilde{t}_2$. Hence it suffices to solve

$$4\tilde{s}_1^3 + 2\tilde{s}_1\tilde{t}_1 = 4\tilde{s}_2^3 + 2\tilde{s}_2\tilde{t}_1 \ \text{ and } \ 3\tilde{s}_1^4 + \tilde{s}_1^2\tilde{t}_1 = 3\tilde{s}_2^4 + \tilde{s}_2^2\tilde{t}_1 \tag{1.4.51}$$

under the condition $\tilde{s}_1 \neq \tilde{s}_2$. The Eq. (1.4.51) can be rewritten as

$$2(\tilde{s}_1^2 + \tilde{s}_1\tilde{s}_2 + \tilde{s}_2^2) + \tilde{t}_1 = 0 \text{ and } (\tilde{s}_1 + \tilde{s}_2)\big(3(\tilde{s}_1^2 + \tilde{s}_2^2) + \tilde{t}_1\big) = 0. \tag{1.4.52}$$

Eliminating $\tilde{t}_1$ from these two equations, we find

$$(\tilde{s}_1 + \tilde{s}_2)(\tilde{s}_1 - \tilde{s}_2)^2 = 0. \tag{1.4.53}$$

Thus we obtain $\tilde{s}_1 + \tilde{s}_2 = 0$ and $\tilde{t}_1 = -2\tilde{s}_1^2$. This implies

$$\begin{cases} x_1(\tilde{s}_1, \tilde{t}_1) = x_1(\tilde{s}_2, \tilde{t}_2) = 0, \\ x_2(\tilde{s}_1, \tilde{t}_1) = x_2(\tilde{s}_2, \tilde{t}_2) = -2\tilde{s}_1^2, \\ y(\tilde{s}_1, \tilde{t}_1) = y(\tilde{s}_2, \tilde{t}_2) = \tilde{s}_1^4 + \dfrac{3}{16}. \end{cases} \tag{1.4.54}$$

In conclusion, every point on the x_2-axis $\{(x_1, x_2) \in \mathbb{C}; x_1 = 0\}$ is a virtual turning point of the Pearcey system $\mathcal{M}_3$.

Remark 1.5 The bicharacteristic strip of $q = \eta^{3/2}q_-$ computed above is geometrically included in the bicharacteristic manifold of $\mathcal{M}_3$. In fact, if we make a change of variables

$$z_1 = x_1, \quad z_2 = x_2 - c(x_1 - 1), \quad \zeta_1 = \xi_1 + c\xi_2, \quad \zeta_2 = \xi_2 \tag{1.4.55}$$

in the description (1.4.44) of the bicharacteristic manifold of $\mathcal{M}_3$, we obtain

$$\begin{aligned} &z_1 = 4\tilde{s}^3 + 2\tilde{s}\tilde{t}, \quad z_2 = \tilde{t} - c(4\tilde{s}^3 + 2\tilde{s}\tilde{t} - 1), \quad y = 3\tilde{s}^4 + \tilde{s}^2\tilde{t} + \frac{3}{16}, \\ &\zeta_1 = c\tilde{s}^2 - \tilde{s}, \quad \zeta_2 = \tilde{s}^2, \quad \eta = 1, \end{aligned} \tag{1.4.56}$$

where $\tilde{s} = s - 1/2$, $\tilde{t} = t - 3/2$. We now restrict (1.4.56) on $z_2 = 0$, that is, we impose

$$\tilde{t} - c(4\tilde{s}^3 + 2\tilde{s}\tilde{t} - 1) = 0, \text{ i.e., } \tilde{t} = \frac{c(4\tilde{s}^3 - 1)}{(1 - 2c\tilde{s})}. \tag{1.4.57}$$

Furthermore, by comparing $\zeta_1 = c\tilde{s}^2 - \tilde{s}$ in (1.4.56) with $\zeta = c^7\tilde{\sigma}^2 - (1/4c)$ in the expression (1.4.15) of the ζ-component of the bicharacteristic strip of q, we relate the two parameters $\tilde{\sigma}$ and $\tilde{s}$ through the relation

$$c^3\tilde{\sigma} = -\big(\tilde{s} - \frac{1}{2c}\big) = -\frac{1}{2c}(2c\tilde{s} - 1). \tag{1.4.58}$$

(Here we take the $-$ sign as we are considering $q = \eta^{3/2}q_-$, not $\eta^{3/2}q_+$.) Then by a straightforward computation we can confirm that the expressions (1.4.20) and (1.4.22) of z and y for the bicharacteristic strip of q coincide with the expressions of z_1 and y in (1.4.56) (except for the constant term of y, which depends on the initial condition) through the relations (1.4.57) and (1.4.58).

1.5 Some Examples of the Stokes Geometry of a System with an Integral Representation of Solutions

In this section we present some intriguing examples of the Stokes geometry of a system with an integral representation of solutions. Here we give only a few numerical results for the Stokes geometry. Mathematical explanations for them will be provided later in Chaps. 2 and 3.

1.5.1 Crossing of Three Ordinary Stokes Curves

First, let us recall the explicit form of the Pearcey system presented in Sect. 1.2:

$$\mathcal{M}_3 : \quad \begin{cases} \eta^{-1}\dfrac{\partial}{\partial x_1}\Psi = P_0\Psi \\ \eta^{-1}\dfrac{\partial}{\partial x_2}\Psi = (Q_0 + \eta^{-1}Q_1)\Psi, \end{cases}$$

where P_0, Q_0 and Q_1 are the following matrices:

$$P_0 = \begin{pmatrix} 0 & 1 & 0 \\ 0 & 0 & 1 \\ -\frac{x_1}{4} & -\frac{x_2}{2} & 0 \end{pmatrix},$$

$$Q_0 = \begin{pmatrix} 0 & 0 & 1 \\ -\frac{x_1}{4} & -\frac{x_2}{2} & 0 \\ 0 & -\frac{x_1}{4} & -\frac{x_2}{2} \end{pmatrix}, \quad Q_1 = \begin{pmatrix} 0 & 0 & 0 \\ 0 & 0 & 0 \\ -\frac{1}{4} & 0 & 0 \end{pmatrix}.$$

We now consider the Stokes geometry of the tangential system of the Pearcey system $\mathcal{M}_3$ to the hyperplane

$$\left\{(x_1, x_2) \in \mathbb{C}^2 \,;\, x_1 = c\right\}.$$

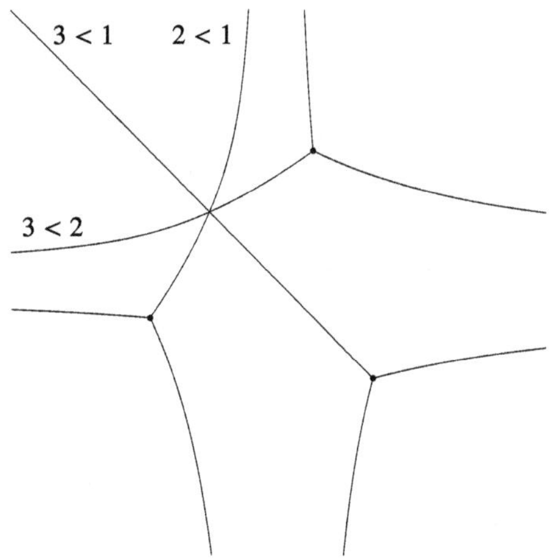

Fig. 1.3 The Stokes geometry of the system $\mathcal{N}_3[x_1 = c]$ for $c = \exp\{(1/8)\pi i\}$

The tangential system is explicitly given by

$$\mathcal{N}_3[x_1 = c]: \quad \eta^{-1}\frac{d}{dx}\Psi = (Q_0 + \eta^{-1}Q_1)\Psi, \tag{1.5.1}$$

where x is used instead of x_2.

A turning point of (1.5.1) is, by definition, a zero of

$$8x^3 + 27c^2.$$

The configuration of turning points and ordinary Stokes curves of $\mathcal{N}[x_1 = c]$ for $c = \exp\{(1/8)\pi i\}$ is shown in Fig. 1.3. We find by numerical computations that, not only for $c = \exp\{(1/8)\pi i\}$ but also for generic values of c, there exists a point where the ordinary Stokes curves emanating from three turning points cross. It is also confirmed numerically that the labels of these three ordinary Stokes curves at the crossing point are $3 < 2$, $2 < 1$ and $3 < 1$. In what follows such a crossing point is called a **triordered-type crossing point**. The triordered-type crossing point of (1.5.1) is stable in the sense that it exists for generic values of the parameter c. Note that there is no virtual turning point for this equation, and hence it is not necessary to consider any new Stokes curve.

1.5.2 *Replacement by a New Stokes Curve for the Crossing of Three Ordinary Stokes Curves*

In the previous subsection we observed the existence of a triordered-type crossing point of three ordinary Stokes curves. In some situations it may happen that one of these three ordinary Stokes curves is replaced by a new Stokes curve. The following equation concretely exemplifies this phenomenon.

Consider the following system of three variables (x_1, x_2, x_3):

$$
\begin{cases}
\left(5\dfrac{\partial^4}{\partial x_1^4} + 3x_3\eta^2\dfrac{\partial^2}{\partial x_1^2} + 2x_2\eta^3\dfrac{\partial}{\partial x_1} + x_1\eta^4\right)\psi = 0 \\
\left(\eta\dfrac{\partial}{\partial x_2} - \dfrac{\partial^2}{\partial x_1^2}\right)\psi = 0 \\
\left(\eta^2\dfrac{\partial}{\partial x_3} - \dfrac{\partial^3}{\partial x_1^3}\right)\psi = 0
\end{cases}
\tag{1.5.2}
$$

which is satisfied by the integral

$$
\psi(x, \eta) = \int \exp\left\{\eta(t^5 + x_3t^3 + x_2t^2 + x_1t)\right\} dt.
$$

Since this integral is related to the singularity of type A_4, we denote (1.5.2) by $\mathscr{M}_4$. The system $\mathscr{M}_4$ is equivalent to

$$
\begin{cases}
\eta^{-1}\dfrac{\partial}{\partial x_1}\Psi = P_0\Psi \\
\eta^{-1}\dfrac{\partial}{\partial x_2}\Psi = (Q_0 + \eta^{-1}Q_1)\Psi \\
\eta^{-1}\dfrac{\partial}{\partial x_3}\Psi = (R_0 + \eta^{-1}R_1)\Psi,
\end{cases}
$$

where P_0, Q_0, Q_1, R_0 and R_1 are given by

$$
P_0 = \begin{pmatrix} 0 & 1 & 0 & 0 \\ 0 & 0 & 1 & 0 \\ 0 & 0 & 0 & 1 \\ -\frac{x_1}{5} & -\frac{2x_2}{5} & -\frac{3x_3}{5} & 0 \end{pmatrix},
$$

$$
Q_0 = \begin{pmatrix} 0 & 0 & 1 & 0 \\ 0 & 0 & 0 & 1 \\ -\frac{x_1}{5} & -\frac{2x_2}{5} & -\frac{3x_3}{5} & 0 \\ 0 & -\frac{x_1}{5} & -\frac{2x_2}{5} & -\frac{3x_3}{5} \end{pmatrix}, \quad
Q_1 = \begin{pmatrix} 0 & 0 & 0 & 0 \\ 0 & 0 & 0 & 0 \\ 0 & 0 & 0 & 0 \\ -\frac{1}{5} & 0 & 0 & 0 \end{pmatrix},
$$

$$
R_0 = \begin{pmatrix} 0 & 0 & 0 & 1 \\ -\frac{x_1}{5} & -\frac{2x_2}{5} & -\frac{3x_3}{5} & 0 \\ 0 & -\frac{x_1}{5} & -\frac{2x_2}{5} & -\frac{3x_3}{5} \\ \frac{3x_1x_3}{25} & \frac{6x_2x_3}{25} & \frac{-5x_1+9x_3^2}{25} & -\frac{2x_2}{5} \end{pmatrix}, \quad
R_1 = \begin{pmatrix} 0 & 0 & 0 & 0 \\ 0 & 0 & 0 & 0 \\ -\frac{1}{5} & 0 & 0 & 0 \\ 0 & -\frac{2}{5} & 0 & 0 \end{pmatrix}.
$$

We consider the Stokes geometry of the tangential system of $\mathscr{M}_4$ to the hyperplane

$$\left\{(x_1, x_2, x_3) \in \mathbb{C}^3 \; ; \; x_1 = c_1, x_3 = c_3\right\},$$

that is,

$$\mathscr{N}_4[x_1 = c_1, x_3 = c_3] : \quad \eta^{-1}\frac{d}{dx}\Psi = \left(Q_0 + \eta^{-1}Q_1\right)\Psi, \tag{1.5.3}$$

where x is used instead of x_2. Again, as in (1.2.8), we denote the restrictions of Q_0 and Q_1 by the same symbols.

A turning point of (1.5.3) is, by definition, a zero of the discriminant of the characteristic equation of Q_0, which explicitly reads as

$$x^4\left\{135x^4 + (-540c_1c_3 + 27c_3^3)x^2 - 400c_1^3 + 360c_1^2c_3^2 - 81c_1c_3^4\right\}.$$

In particular, a turning point $x = 0$ is a non-hereditary turning point. Figure 1.4 shows the Stokes geometry of $\mathscr{N}_4[x_1 = c_1, x_3 = c_3]$ for $c_1 = 6/5 + i$, $c_3 = -1 - i$. Since the non-hereditary turning point is redundant as verified in Sect. 1.2, it is omitted in Fig. 1.4. Similarly to the case of $\mathscr{N}_3[x_1 = c]$, the system $\mathscr{N}_4[x_1 = c_1, x_3 = c_3]$ for $c_1 = 6/5 + i$, $c_3 = -1 - i$ has also a triordered-type crossing point p of three ordinary Stokes curves (cf. Fig. 1.4). However, changing the parameter c_3 continuously, we observe that p becomes an intersection point of two ordinary Stokes curves and one new Stokes curve, as is shown in Fig. 1.5. That is, one of the three ordinary Stokes curves constituting the triordered-type crossing point p is replaced by a new Stokes curve. In the Appendix, we discuss the Stokes phenomenon related to this replacement.

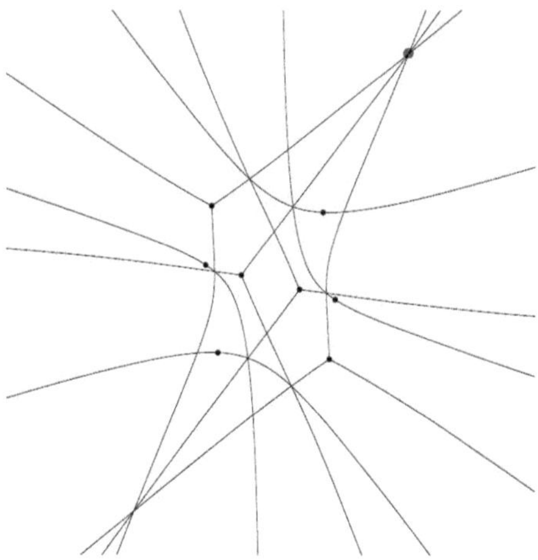

Fig. 1.4 The Stokes geometry of the system $\mathscr{N}_4[x_1 = c_1, x_3 = c_3]$ for $c_1 = 6/5 + i$, $c_3 = -1 - i$. The *red point* is the crossing point p in question

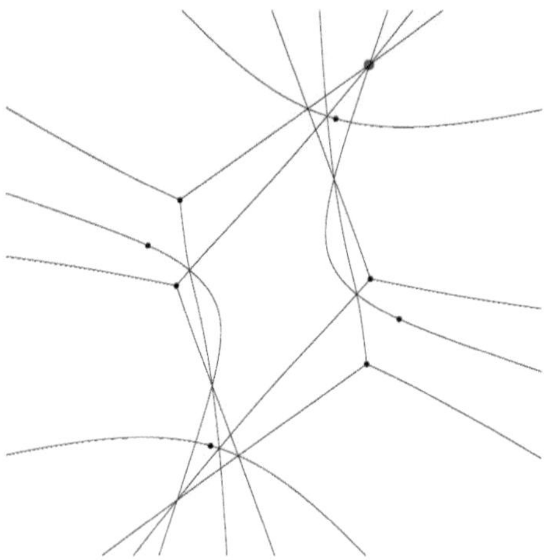

Fig. 1.5 The Stokes geometry of the system $\mathcal{N}_4[x_1 = c_1, x_3 = c_3]$ for $c_1 = 6/5 + i, c_3 = -i$. The *red point* is the crossing point p in question

1.5.3 Crossing of Three Stokes Curves and Non-hereditary Turning Points

In Sects. 1.5.1 and 1.5.2 we observe that triordered-type crossing phenomena of three Stokes curves occur with ordinary Stokes curves and new Stokes curves. These crossing phenomena also occur with Stokes curves emanating from a non-hereditary turning point. In this subsection we show some intriguing interplay between the crossing phenomena of Stokes curves and non-hereditary turning points by using the system $\mathcal{N}(c)$ discussed in Sect. 1.2.

The system $\mathcal{N}(c)$ is the tangential system of the Pearcey system $\mathcal{M}_3$ to the hyperplane

$$\left\{(x_1, x_2) \in \mathbb{C}^2 \; ; \; x_2 = c(x_1 - 1)\right\}$$

and explicitly given by

$$\mathcal{N}(c) : \quad \eta^{-1}\frac{d}{dz}\Psi = ((P_0 + cQ_0) + \eta^{-1}cQ_1)\Psi. \tag{1.5.4}$$

Here x_1 and x_2 appearing in the components of the matrices P_0 and Q_0 are replaced by $x_1 = z$ and $x_2 = c(z - 1)$, respectively. The restrictions of the matrices P_0, Q_0 and Q_1 are denoted by the same symbols as before.

A turning point of (1.5.4) is, by definition, a zero of

$$f_1(z) f_2(z)^2,$$

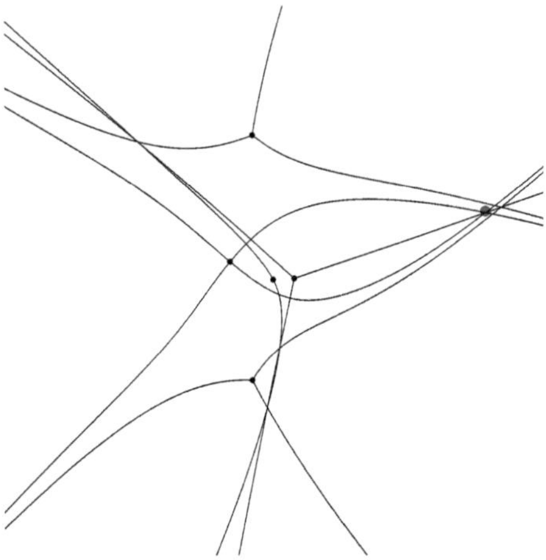

Fig. 1.6 The Stokes geometry of the system $\mathcal{N}(c)$ for $c = \exp\{(22/32)\pi i\}$. The *red point* is the crossing point p in question

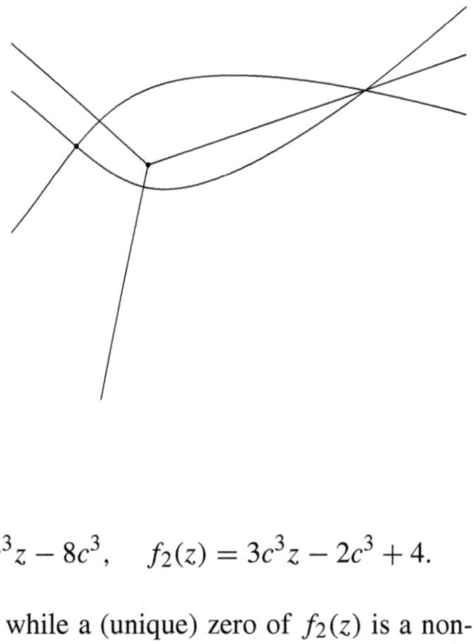

Fig. 1.7 Leaf-type configuration, which is obtained by enlarging relevant parts of Fig. 1.6

where

$$f_1(z) = 8c^3z^3 + (27 - 24c^3)z^2 + 24c^3z - 8c^3, \quad f_2(z) = 3c^3z - 2c^3 + 4.$$

Zeros of $f_1(z)$ are simple turning points, while a (unique) zero of $f_2(z)$ is a non-hereditary double turning point. In Fig. 1.6 three Stokes curves cross at a point p and two of them are Stokes curves emanating from the non-hereditary double turning point. Furthermore, the region surrounded by these two Stokes curves contains another Stokes curve constituting the crossing point p and a simple turning point of its starting point. Such a configuration of Stokes curves is called a **leaf-type configuration** in what follows. Figure 1.7 is an enlarged version of Fig. 1.6 depicting only the parts related to a leaf-type configuration. As will be verified in Sect. 3.4, a crossing point of three Stokes curves in a leaf-type configuration is always a triordered-type crossing point.

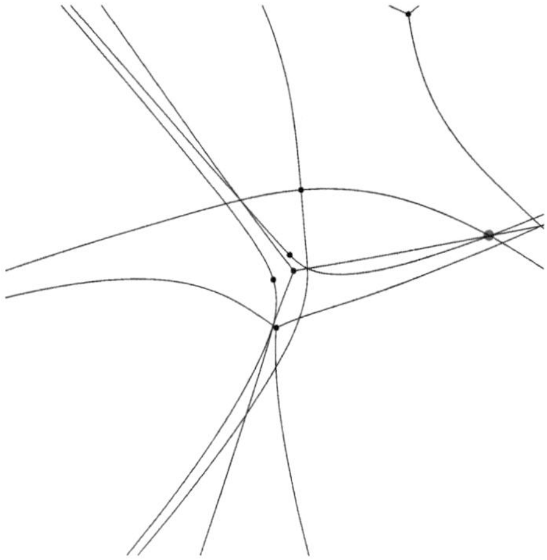

Fig. 1.8 The Stokes geometry of the system $\mathcal{N}(c)$ for $c = \exp\{(26/32)\pi i\}$. The *red point* is the crossing point p in question

Similarly to the discussion of Sect. 1.5.2, one of the three Stokes curves crossing at the point p is replaced by a new Stokes curve when the parameter c is changed, as shown in Fig. 1.8. The activeness of the Stokes curves related to this replacement will be discussed in Sect. 3.2 and the Appendix. Note that this new Stokes curve in Fig. 1.8 emanates from a virtual turning point which originates from the non-hereditary double turning point. Otherwise stated, the new Stokes curve and the virtual turning point in question become relevant only when we consider a perturbed system of $\mathcal{N}(c)$. One peculiar point of this virtual turning point is that it is related to the periodic structure of $\mathcal{N}(c)$, which is defined through the so-called **bicharacteristic chain**. Note that singularities of solutions of the Borel transformed equation propagate along the bicharacteristic chain when a double turning point exists. In the case of $\mathcal{N}(c)$ the periodic structure is explicitly described by a complex number called the **period**

$$\varpi = -\frac{1}{c^4}\left(\frac{2c^3-1}{3}\right)^{3/2}.$$

Here, in particular, we consider the case $c = 1$. In this case the period ϖ is real and the Stokes geometry of $\mathcal{N}(c)$ is shown in Fig. 1.9, where we observe that three simple turning points seem to be connected with the non-hereditary double turning point. The detailed discussion for the period will be given in Chap. 2 and the period of $\mathcal{N}(c)$ will be explained in detail in Chap. 3. We also note that the examples $\mathcal{N}_3[x_1 = c]$ and $\mathcal{N}_4[x_1 = c_1, x_3 = c_3]$ previously discussed do not have such a periodic structure.

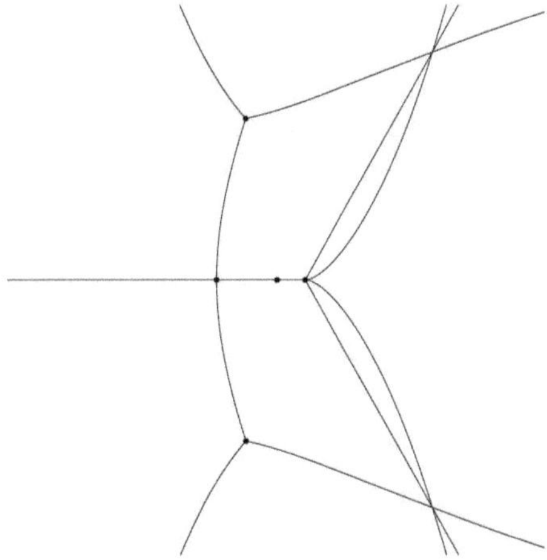

Fig. 1.9 The Stokes geometry of the system $\mathcal{N}(c)$ for $c = 1$

1.5.4 Trilemma-Type Crossing of Three Stokes Curves

In the previous subsections we have seen several examples of triordered-type crossing points of three Stokes curves. In this subsection we discuss a different type of crossing points of Stokes curves.

We consider

$$\begin{cases} \left(3\dfrac{\partial^3}{\partial x_1^3} + 2x_2\eta\dfrac{\partial^2}{\partial x_1^2} + x_1\eta^2\dfrac{\partial}{\partial x_1} - \alpha\eta^3\right)\psi = 0 \\ \left(\eta\dfrac{\partial}{\partial x_2} - \dfrac{\partial^2}{\partial x_1^2}\right)\psi = 0 \end{cases} \tag{1.5.5}$$

which belongs to the class of **hypergeometric systems of two variables** studied in [OK]. Since this system has a solution given by the integral

$$\psi(x,\eta) = \int \exp\left\{\eta(t^3 + x_2t^2 + x_1t)\right\} t^{-\alpha\eta-1}dt$$

and this integral is determined by the partition $(1, 4)$ of the natural number 5, (1.5.5) is called the $(1, 4)$ **hypergeometric system** and denoted by $\mathcal{M}_{(1,4)HG}$.

The system $\mathcal{M}_{(1,4)HG}$ is equivalent to

$$\mathcal{M}_{(1,4)HG} : \begin{cases} \eta^{-1}\dfrac{\partial}{\partial x_1}\Psi = P_0\Psi \\ \eta^{-1}\dfrac{\partial}{\partial x_2}\Psi = (Q_0 + \eta^{-1}Q_1)\Psi, \end{cases}$$

where P_0, Q_0 and Q_1 are given by

$$P_0 = \begin{pmatrix} 0 & 1 & 0 \\ 0 & 0 & 1 \\ \frac{\alpha}{3} & -\frac{x_1}{3} & -\frac{2x_2}{3} \end{pmatrix},$$

$$Q_0 = \begin{pmatrix} 0 & 0 & 1 \\ \frac{\alpha}{3} & -\frac{x_1}{3} & -\frac{2x_2}{3} \\ -\frac{2\alpha x_2}{9} & \frac{2x_1x_2+3\alpha}{9} & \frac{-3x_1+4x_2^2}{9} \end{pmatrix}, \quad Q_1 = \begin{pmatrix} 0 & 0 & 0 \\ 0 & 0 & 0 \\ 0 & -\frac{1}{3} & 0 \end{pmatrix}.$$

We consider the Stokes geometry of the tangential system of the $(1,4)$ hypergeometric system $\mathscr{M}_{(1,4)HG}$ to the hyperplane

$$\left\{(x_1, x_2) \in \mathbb{C}^2 \,;\, x_2 = c\right\},$$

that is,

$$\mathscr{N}_{(1,4)HG}[x_2 = c] : \quad \eta^{-1}\frac{d}{dx}\Psi = P_0\Psi, \tag{1.5.6}$$

where x is used instead of x_1 and the restriction of P_0 is denoted by the same symbol.

A turning point of (1.5.6) is, by definition, a zero of

$$12x^3 - 4c^2x^2 + 108\alpha cx + 243\alpha^2 - 32\alpha c^3.$$

As in the previous examples, the system $\mathscr{N}_{(1,4)HG}[x_2 = c]$ has also a crossing point p of three ordinary Stokes curves. See Fig. 1.10. However, unlike previous

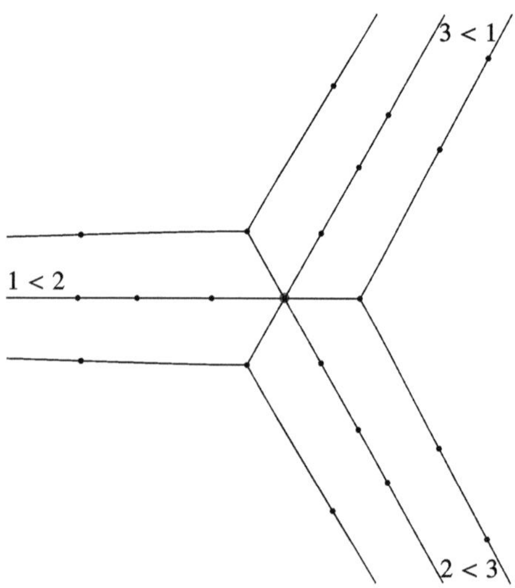

Fig. 1.10 The Stokes geometry of the system $\mathscr{N}_{(1,4)HG}[x_2 = c]$ for $c = 0$ and $\alpha = -i$. The *red point* is the crossing point p in question

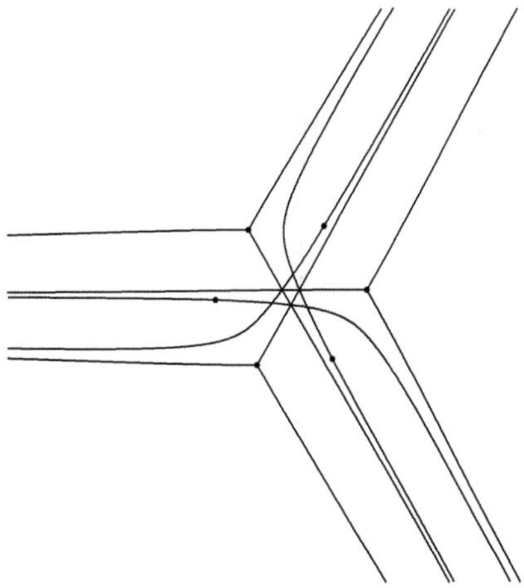

Fig. 1.11 The Stokes geometry of the system $\mathcal{N}_{(1,4)HG}[x_2 = c]$ for $c = 0$ and $\alpha = 1/10 - i$

examples, this crossing point p is not a triordered-type crossing point. In fact, we can numerically confirm that the labels of three Stokes curves crossing at p satisfy the trilemma relationship, i.e., $1 < 2, 2 < 3, 3 < 1$. In what follows such a crossing point of three Stokes curves is called a **trilemma-type crossing point**. Furthermore, as is shown in Fig. 1.11, this crossing point of three Stokes curves splits into three crossing points of two Stokes curves when parameters of the system are changed. This makes a clear contrast to triordered-type crossing points.

Unlike the previous examples $\mathcal{M}_3$ and $\mathcal{M}_4$, the system $\mathcal{M}_{(1,4)HG}$ has a periodic structure described by the period $\varpi = 2\pi i\alpha$. Therefore, $\mathcal{N}_{(1,4)HG}[x_2 = c]$ also has a period ϖ and this period creates an infinite number of virtual turning points. The points on Stokes curves in Fig. 1.10 depict some of these virtual turning points. In Fig. 1.11 only the virtual turning points related to the three crossing points are depicted. Note that the period ϖ in the case of Fig. 1.10 satisfies $\Im\varpi = 0$, and a trilemma-type crossing point only appears when $\Im\varpi = 0$.

1.5.5 Some Other Degenerate Configurations of the Stokes Geometry

At the end of this section we give some other examples of degenerate configurations of the Stokes geometry.

We consider the Stokes geometry of the tangential system of the (1, 4) hypergeometric system $\mathcal{M}_{(1,4)HG}$ to the hyperplane

$$\left\{(x_1, x_2) \in \mathbb{C}^2 \,;\; x_1 = c\right\},$$

that is,

$$\mathscr{N}_{(1,4)HG}[x_1 = c]: \quad \eta^{-1}\frac{d}{dx}\Psi = (Q_0 + \eta^{-1}Q_1)\Psi, \tag{1.5.7}$$

where x is used instead of x_2 and the restrictions of Q_0 and Q_1 are denoted by the same symbols again.

A turning point of (1.5.7) is, by definition, a zero of

$$(2cx + 3\alpha)^2(32\alpha x^3 + 4c^2x^2 - 108\alpha cx - 243\alpha^2 - 12c^3).$$

For $c \neq 0$ $\mathscr{N}_{(1,4)HG}[x_1 = c]$ has a non-hereditary double turning point at $x = -(3\alpha)/(2c)$. Furthermore, in the sense similar to the **simple-pole-type turning points** of second-order differential equations discussed in [Ko], $\mathscr{N}_{(1,4)HG}[x_1 = c]$ has a simple pole at $x = \infty$. That is, by using the coordinate transformation $z = 1/x$, we can confirm that $\mathscr{N}_{(1,4)HG}[x_1 = c]$ has two characteristic roots $\zeta_\pm$ with the following property at $z = 0$:

$$(\zeta_+ - \zeta_-)\,dx = \left\{-c\left(\frac{\alpha}{2}\right)^{1/2} z^{-1/2} + \cdots\right\}dz.$$

(The third characteristic root is non-ramified and separated from $\zeta_\pm$ at $z = 0$. In fact, it has a pole at $z = 0$.) The double turning point and simple pole merge when $c \to 0$, and $\mathscr{N}_{(1,4)HG}[x_1 = 0]$ has a simple turning point at $x = \infty$.

As explained at the end of the previous subsection, $\mathscr{M}_{(1,4)HG}$ has a period $\varpi = 2\pi i\alpha$ and consequently $\mathscr{N}_{(1,4)HG}[x_1 = c]$ also has a period ϖ. Furthermore, $\mathscr{N}_{(1,4)HG}[x_1 = c]$ has another period $\varpi' = \pi i\alpha - 4\,(-c/3)^{3/2}$ that originates from the existence of the non-hereditary double turning point. This new period ϖ' is expected to create a different set of infinitely many virtual turning points, and they are redundant. In this subsection, forgetting all virtual turning points and new Stokes curves emanating from them, we only consider Stokes curves emanating from ordinary turning points of $\mathscr{N}_{(1,4)HG}[x_1 = c]$ for several values of (c, α).

First, we consider the case $(c, \alpha) = (0, 2i)$. The Stokes geometry is shown in Fig. 1.12. In this case we have $\varpi = -2\pi$ and $\varpi' = -\pi$, which imply the periods are real and linearly dependent over $\mathbb{Z}$. As its consequence, three simple turning points are connected with the simple turning point at $x = \infty$ by Stokes curves in Fig. 1.12.

We next consider the case $(c, \alpha) = (-3/4, 2i)$. In this case we find $\varpi = -4\pi$ and $\varpi' = -2\pi - 1/2$; the periods ϖ and ϖ' are real and linearly independent over $\mathbb{Z}$. The Stokes geometry is shown in Fig. 1.13, where we observe that three simple turning points as well as the simple pole at $x = \infty$ are connected with the non-hereditary double turning point by Stokes curves.

The degeneration observed in Figs. 1.12 and 1.13 are resolved if the parameters (c, α) are further changed so that ϖ and ϖ' may become non-real numbers. For example, Fig. 1.14 shows the Stokes geometry for $(c, \alpha) = (-3/4, 2i) +$

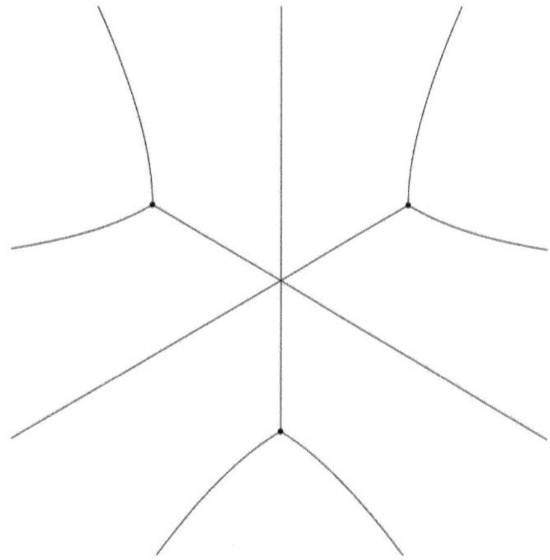

Fig. 1.12 The Stokes geometry of $\mathcal{N}_{(1,4)HG}[x_1 = c]$ for $(c, \alpha) = (0, 2i)$

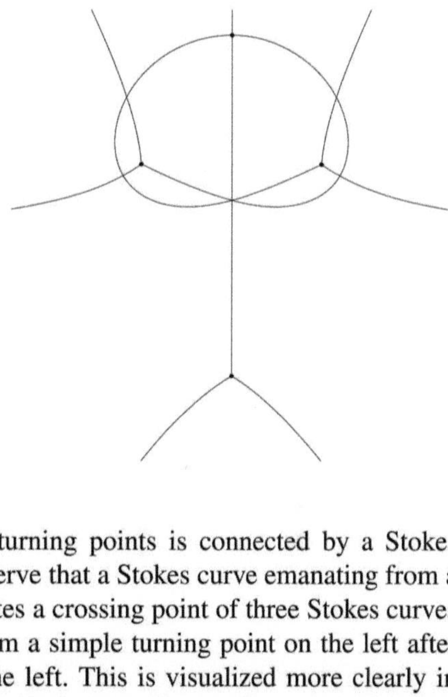

Fig. 1.13 The Stokes geometry of $\mathcal{N}_{(1,4)HG}[x_1 = c]$ for $(c, \alpha) = (-3/4, 2i)$

$((3/50)i, -(1/100))$, where no pair of turning points is connected by a Stokes curve. Note that, in Fig. 1.14, we can observe that a Stokes curve emanating from a simple turning point on the right constitutes a crossing point of three Stokes curves with another Stokes curve emanating from a simple turning point on the left after encircling the simple turning point on the left. This is visualized more clearly in Fig. 1.15, an enlarged version of Fig. 1.14. Such a peculiar configuration of Stokes curves is referred to as a **tadpole-type configuration** in what follows. The tadpole-type configuration is often encountered.

The case where only one period is real can be considered as well. More detailed discussions including such a case with one real period will be done in Chap. 3.

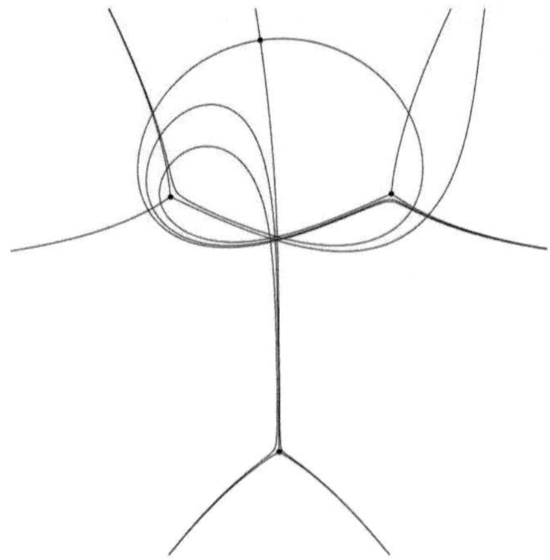

Fig. 1.14 The Stokes geometry of $\mathscr{N}_{(1,4)HG}[x_1 = c]$ for $(c, \alpha) = (-3/4, 2i) + ((3/50)i, -(1/100))$

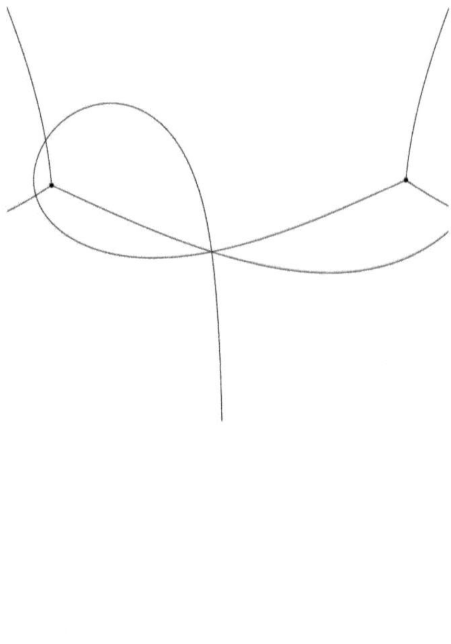

Fig. 1.15 Tadpole-type configuration, which is obtained by enlarging relevant parts of Fig. 1.14

Chapter 2
Degeneration of the Stokes Geometry for Higher-Order Equations

Abstract In Chap. 1 we observed that there are two kinds of crossing points of three Stokes curves, that is, trilemma-type crossing points and triordered-type crossing points. Although the configuration of Stokes curves are similar between these two kinds of crossing points, they have quite different characters. In this chapter we discuss their difference from the viewpoint of the degeneration of the Stokes geometry. As explained in Honda et al. (Virtual Turning Points. Springer Briefs in Mathematical Physics, vol. 4. Springer-Verlag, Berlin, 2015, Chapter 2), degeneration of the Stokes geometry for higher-order equations is described by "growing trees" and a trilemma-type crossing point is the simplest example of such growing trees. Furthermore, degeneration of the Stokes geometry is related to the location of singularities of Borel transformed WKB solutions via the so-called periods. Using the $(1, 4)$ hypergeometric system discussed in Sect. 1.5.4 as an example, we will confirm this intriguing fact by investigating its bicharacteristic curve concretely. In contrast with this, a triordered-type crossing point has no relationship with degeneration of the Stokes geometry.

2.1 What is a New Feature of Degeneration of the Stokes Geometry for Higher-Order Equations?

In Chap. 1 we have presented several intriguing examples of the Stokes geometry of systems with integral representation of solutions. In particular, in Sect. 1.5 we observed that there occur two types of crossing phenomena of three Stokes curves, that is, triordered-type crossing points and trilemma-type crossing points. Such crossing phenomena of Stokes curves are related to the **degeneration of the Stokes geometry** for higher-order ordinary differential equations. Taking these crossing phenomena of Stokes curves into account, we discuss degeneration of the Stokes geometry for higher-order equations and its relation to the location of singularities of relevant Borel transformed WKB solutions in this chapter.

Degeneration of the Stokes geometry for higher-order equations is quite different from that for second-order equations and has several new features. For example, a trilemma-type crossing of three Stokes curves is a typical degeneration of the Stokes

S. Hirose et al., *Virtual Turning Points II*, SpringerBriefs in Mathematical Physics 52, https://doi.org/10.1007/978-981-95-2817-2_2

geometry which is peculiar to higher-order equations. More generally, the **growing trees** discussed in [VTP, Chapter 2] are nothing but the theory of degeneration of the Stokes geometry for higher-order equations. Note that a trilemma-type crossing of three Stokes curves is the simplest example of **bidirectional binary trees** (cf. [VTP, Definition 2.4.3] for the definition of bidirectional binary trees). On the other hand, as was discussed in [VTP, Section 2.4], a trilemma-type crossing point is generated from a Stokes segment connecting two turning points of the same type, a typical degenerate configuration of the Stokes geometry for second-order equations, via the so-called **bifurcation phenomena of Stokes curves**. Thus the degeneration of the Stokes geometry for second-order equations can be regarded as the origin of the degeneration for higher-order equations. Hence, before discussing degeneration of the Stokes geometry for higher-order equations, we first consider the case of second-order equations in this section.

We explain the case of second-order equations by taking a particular example. Let

$$\left(\frac{d^2}{dx^2} - \eta^2 Q_0(x,\lambda)\right)\psi = 0 \quad \text{with} \quad Q_0(x,\lambda) = \frac{x^2}{4} - \lambda \tag{2.1.1}$$

be the **Weber equation** with a large parameter $\eta > 0$, where λ is assumed to be a non-zero complex constant. Equation (2.1.1) has two simple turning points at $x = \pm 2\sqrt{\lambda}$ and let ϖ be the contour integral of $\sqrt{Q_0(x,\lambda)}\,dx$ defined by

$$\varpi = 2\int_{-2\sqrt{\lambda}}^{2\sqrt{\lambda}} \sqrt{Q_0(x,\lambda)}\,dx = \int_{\gamma_0} \sqrt{Q_0(x,\lambda)}\,dx, \tag{2.1.2}$$

where γ_0 denotes a contour on the Riemann surface of $\sqrt{Q_0}$ shown in Fig. 2.1. When ϖ is a real number, two turning points $\pm 2\sqrt{\lambda}$ are connected by a Stokes curve $\operatorname{Im} \displaystyle\int_{-2\sqrt{\lambda}}^{x} \sqrt{Q_0}\,dx = 0$ and the Stokes geometry of (2.1.1) becomes degenerate (cf. Fig. 2.2). Note that, in the case of (2.1.1), $\sqrt{Q_0}\,dx$ has a singularity at $x = \infty$ and the residue of $\sqrt{Q_0}\,dx$ at $x = \infty$ is equal to $\pm\lambda$. (The choice of the sign depends on the branch of $\sqrt{Q_0}$.) This is readily confirmed by using a change of variable $x = 1/w$ and observing

$$\begin{aligned}\sqrt{Q_0}\,dx &= \sqrt{\frac{x^2}{4} - \lambda}\,dx = \sqrt{\frac{1}{4w^2} - \lambda}\left(-\frac{1}{w^2}\right)dw \\ &= \mp\frac{1}{2w^3}(1 - 4\lambda w^2)^{1/2}dw = \mp\frac{1}{2w^3}(1 - 2\lambda w^2 + \cdots)\,dw. \end{aligned}\tag{2.1.3}$$

In fact, it follows from (2.1.3) that

$$\operatorname*{Res}_{x=\infty} \sqrt{Q_0}\,dx = \operatorname*{Res}_{w=0} \mp\frac{1}{2w^3}(1 - 2\lambda w^2 + \cdots)\,dw = \pm\lambda.$$

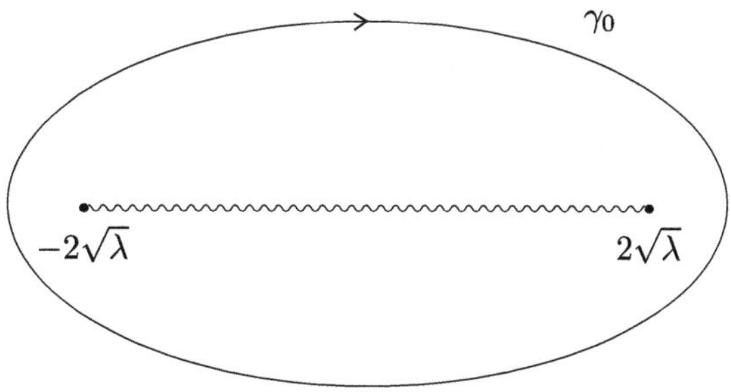

Fig. 2.1 Contour γ_0 (A wiggly line designates a cut to define $\sqrt{Q_0(x,\lambda)}$.)

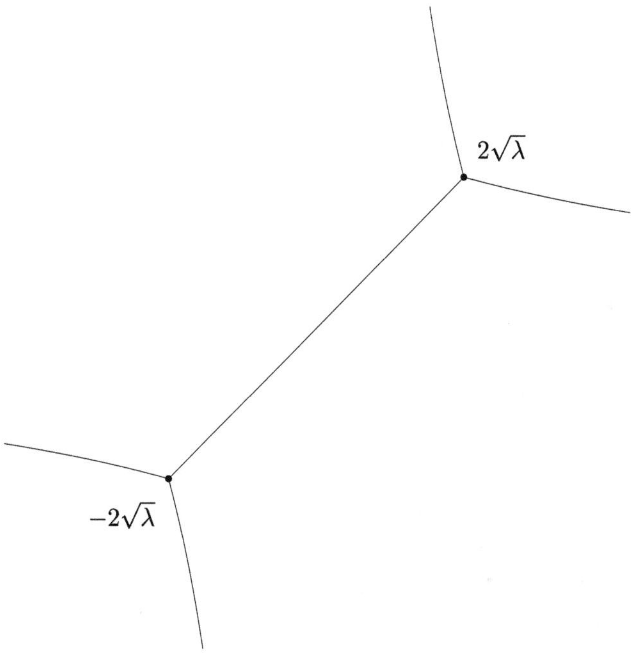

Fig. 2.2 Stokes geometry of (2.1.1) when $\varpi \in \mathbb{R}$, i.e., $\lambda \in i\mathbb{R}$

Hence ϖ coincides with $2\pi i\lambda$ (up to the sign). The value of a contour integral of $\sqrt{Q_0}\,dx$ along a closed path on the Riemann surface of $\sqrt{Q_0}$ like $\varpi = 2\pi i\lambda$ is called a **period** of (2.1.1). Thus the set of periods of Eq. (2.1.1) is equal to $(2\pi i\lambda)\mathbb{Z}$, i.e., integral multiples of $2\pi i\lambda$. The degeneration of the Stokes geometry of (2.1.1) is controlled by the period $\varpi = 2\pi i\lambda$.

On the other hand, the period $\varpi = 2\pi i\lambda$ is related also to the location of singularities of Borel transformed WKB solutions of (2.1.1). To see this fact, let us investigate the bicharacteristic curve of

$$\left(\frac{\partial^2}{\partial x^2} - Q_0(x,\lambda)\frac{\partial^2}{\partial y^2}\right)\psi_B = 0, \tag{2.1.4}$$

that is, the Borel transform of (2.1.1), since the Borel transform of a WKB solution satisfies (2.1.4) as was explained in Sect. 1.1 and the singularities of a solution of (2.1.4) propagate along its bicharacteristic curve. Now, letting

$$\sigma(x, \xi, \eta) = \xi^2 - Q_0(x, \lambda)\eta^2 = \xi^2 - \left(\frac{x^2}{4} - \lambda\right)\eta^2$$

denote the principal symbol of (2.1.4), we find a bicharacteristic strip of (2.1.4) is a solution of

$$\begin{cases} \dfrac{dx}{dt} = \dfrac{\partial\sigma}{\partial\xi} = 2\xi, \\ \dfrac{dy}{dt} = \dfrac{\partial\sigma}{\partial\eta} = -2\left(\dfrac{x^2}{4} - \lambda\right)\eta, \\ \dfrac{d\xi}{dt} = -\dfrac{\partial\sigma}{\partial x} = \dfrac{x}{2}\eta^2, \\ \dfrac{d\eta}{dt} = -\dfrac{\partial\sigma}{\partial y} = 0, \end{cases} \tag{2.1.5}$$

which satisfies $\sigma(x(t), \xi(t), \eta(t)) = 0$. A bicharacteristic curve is its projection to the base manifold $\mathbb{C}^2_{(x,y)}$, i.e., the (x, y)-component $(x(t), y(t))$. Taking $(x(0), y(0), \xi(0), \eta(0)) = (2\sqrt{\lambda}, 0, 0, 1)$ as an initial condition, we seek for a solution of (2.1.5) explicitly. First, we find $\eta(t)$ identically equals 1 by the fourth equation of (2.1.5). Then the first and the third equations become

$$\frac{dx}{dt} = 2\xi, \quad \frac{d\xi}{dt} = \frac{x}{2},$$

which imply

$$x(t) = \sqrt{\lambda}(e^t + e^{-t}), \quad \xi(t) = \frac{\sqrt{\lambda}}{2}(e^t - e^{-t})$$

in view of the initial condition $(x(0), \xi(0)) = (2\sqrt{\lambda}, 0)$. Substituting these results into the second equation, we obtain

$$\frac{dy}{dt} = -\frac{\lambda}{2}(e^{2t} - 2 + e^{-2t}).$$

Hence we have

$$y(t) = \lambda\left(-\frac{1}{4}e^{2t} + t + \frac{1}{4}e^{-2t}\right).$$

Thus the bicharacteristic curve of (2.1.4) is explicitly described by

$$\begin{cases} x(t) = \sqrt{\lambda}(e^t + e^{-t}), \\ y(t) = \lambda\left(-\frac{1}{4}e^{2t} + t + \frac{1}{4}e^{-2t}\right). \end{cases} \tag{2.1.6}$$

If we substitute solutions (with respect to t) of an equation $x(t) = x$ for each fixed x into the second equation of (2.1.6), we obtain singularities $y = y(x)$ of the Borel transform of WKB solutions in the y-plane, i.e., the Borel plane. It follows from the first equation of (2.1.6) that solutions of $x(t) = x$ are given by

$$e^t = \frac{x}{2\sqrt{\lambda}} \mp \sqrt{\frac{x^2}{4\lambda} - 1},$$

that is,

$$t = \log\left(\frac{x}{2\sqrt{\lambda}} \mp \sqrt{\frac{x^2}{4\lambda} - 1}\right).$$

Hence we finally obtain

$$\begin{aligned} y(x) &= \lambda\left[\pm\frac{x}{2\sqrt{\lambda}}\sqrt{\frac{x^2}{4\lambda} - 1} + \log\left(\frac{x}{2\sqrt{\lambda}} \mp \sqrt{\frac{x^2}{4\lambda} - 1}\right)\right] \\ &= \lambda\left[\pm\frac{x}{2\sqrt{\lambda}}\sqrt{\frac{x^2}{4\lambda} - 1} - \log\left(\frac{x}{2\sqrt{\lambda}} \pm \sqrt{\frac{x^2}{4\lambda} - 1}\right)\right]. \end{aligned} \tag{2.1.7}$$

Note that the right-hand side of (2.1.7) is equal to $\pm\int_{2\sqrt{\lambda}}^{x} \sqrt{Q_0}\,dx$, which is confirmed just by differentiating (2.1.7) with respect to x. Formula (2.1.7) describes the location of singularities of Borel transformed WKB solutions of (2.1.1). In view of the multivaluedness of the logarithmic function, (2.1.7) can be expressed more precisely as

$$y(x) = \lambda\left[\pm\frac{x}{2\sqrt{\lambda}}\sqrt{\frac{x^2}{4\lambda} - 1} - \log\left(\frac{x}{2\sqrt{\lambda}} \pm \sqrt{\frac{x^2}{4\lambda} - 1}\right)\right] + 2\pi i\lambda n \tag{2.1.8}$$

with an integer n. Thus the periods $2\pi i\lambda\mathbb{Z}$ appear also in the description of the location of singularities of Borel transformed WKB solutions; singularities of Borel transformed WKB solutions have a periodic structure with the period $\varpi = 2\pi i\lambda$.

The above computation of the bicharacteristic curve tells us that all the singularities with this periodic structure are cognate and they are all connected by the time variable t of the bicharacteristic strip. As a consequence of this periodic structure of the singularities, we find that the Borel summability of WKB solutions of (2.1.1) breaks down when the period $\varpi = 2\pi i\lambda$ is real, that is, when the Stokes geometry of (2.1.1) becomes degenerate.

In subsequent sections we will discuss to what extent the relation between degeneration of the Stokes geometry and the location of singularities of Borel transformed WKB solutions via the periods can be extended to higher-order equations.

2.2 Genesis of the Theory of Growing Trees; an Initiating Observation of Shunsuke Sasaki

In the preceding section we mentioned that a trilemma-type crossing of three Stokes curves observed in Sect. 1.5 is the simplest example of (bidirectional) binary trees for higher-order equations and that a trilemma-type crossing point is generated from a Stokes segment, that is, a segment of a Stokes curve connecting two turning points of the same type, via the so-called bifurcation phenomena of Stokes curves. Such an intriguing fact was first observed by Shunsuke Sasaki in his master thesis (cf. [S1], [S2]). In this section we explain the observation of Shunsuke Sasaki by using his original example, which is discussed in [S1] (and also in [AHKKoNSST], [AKKoST] and [AKSST]) and simpler than the examples dealt with in [VTP, Chapter 2], of a system of first-order linear differential equations associated with the **Noumi–Yamada system**.

As explained in [VTP, Chapter 2], the Noumi–Yamada systems ([NY]) are higher-order versions of the classical fourth and fifth **Painlevé equations**, the first member of which is of the following form:

$$\eta^{-1}\frac{du_j}{dt} = u_j(u_{j+1} - u_{j+2}) + \alpha_j \quad (j = 0, 1, 2), \tag{2.2.1}$$

where u_j are unknown functions and α_j are constants satisfying the constraint $\alpha_0 + \alpha_1 + \alpha_2 = \eta^{-1}$. Here we assume a cyclic relation $u_{j+3} = u_j$ and also assume that $u_0 + u_1 + u_2 = t$ holds. (The latter relation is consistent with the above constraint $\alpha_0 + \alpha_1 + \alpha_2 = \eta^{-1}$.) Note that (2.2.1) can be considered to be a symmetric form of the fourth Painlevé equation. The nonlinear Eq. (2.2.1) describes the compatibility condition of the following system of first-order linear differential equations in the variables x and t:

$$-\eta^{-1}x\frac{\partial}{\partial x}\begin{pmatrix}\psi_0\\ \psi_1\\ \psi_2\end{pmatrix}=\begin{pmatrix}(2\alpha_1+\alpha_2)/3 & u_1 & 1\\ x & (-\alpha_1+\alpha_2)/3 & u_2\\ xu_0 & x & -(\alpha_1+2\alpha_2)/3\end{pmatrix}\begin{pmatrix}\psi_0\\ \psi_1\\ \psi_2\end{pmatrix}, \tag{2.2.2}$$

$$\eta^{-1}\frac{\partial}{\partial t}\begin{pmatrix}\psi_0\\ \psi_1\\ \psi_2\end{pmatrix}=\begin{pmatrix}u_2-t/2 & -1 & 0\\ 0 & u_0-t/2 & -1\\ -x & 0 & u_1-t/2\end{pmatrix}\begin{pmatrix}\psi_0\\ \psi_1\\ \psi_2\end{pmatrix}. \tag{2.2.3}$$

Here we mainly consider the first Eq. (2.2.2) of this system. It is proved in [T4] that, if a 0-**parameter solution** (i.e., formal power series solution in η^{-1}) of (2.2.1) is substituted into the coefficients of (2.2.2) and (2.2.3), a Stokes curve connects two turning points of (2.2.2) when t lies on a Stokes curve emanating from a turning point of (2.2.1) provided that t is sufficiently close to the turning point. Note that for a system of linear differential equations of the form

$$\eta^{-1}\frac{d}{dx}\psi=\big(A_0(x)+\eta^{-1}A_1(x)+\eta^{-2}A_2(x)+\cdots\big)\psi, \tag{2.2.4}$$

where ψ is an unknown vector, a **turning point** and a **Stokes curve** are defined in a manner similar to the case of higher-order ordinary differential Eq. (1.1.1) by considering the characteristic equation $p(x,\zeta)=\det(\zeta I-A_0(x))$ of the top-order part $A_0(x)$ of the coefficients of (2.2.4) as a substitute for (1.1.5). (On the other hand, see [VTP, Chapter 2] for the definition of a turning point and a Stokes curve of the nonlinear Eq. (2.2.1).) For example, take $\alpha_0=1+0.6i$, $\alpha_1=0.2-0.1i$, and let $t_1=-1.6104-0.2268i$, which lies on a Stokes curve of (2.2.1) emanating from a turning point $t^*=-1.6276-0.0986i$ (cf. Fig. 2.3). Then the configuration of Stokes curves of (2.2.2) at $t=t_1$ becomes as indicated in Fig. 2.4a. The linear system (2.2.2) has one double turning point d and two simple turning points s_1 and s_2, and two turning points d and s_1 among the three are connected by a Stokes curve at $t=t_1$. Thus at a point $t=t_1$ on a Stokes curve of (2.2.1) a degenerate configuration is observed with the associated linear system (2.2.2). However, if we

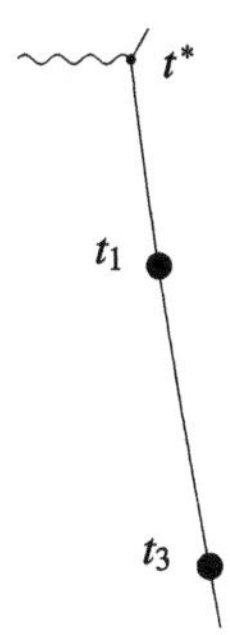

Fig. 2.3 Stokes curve of (2.2.1) emanating from $t^*=-1.6276-0.0986i$

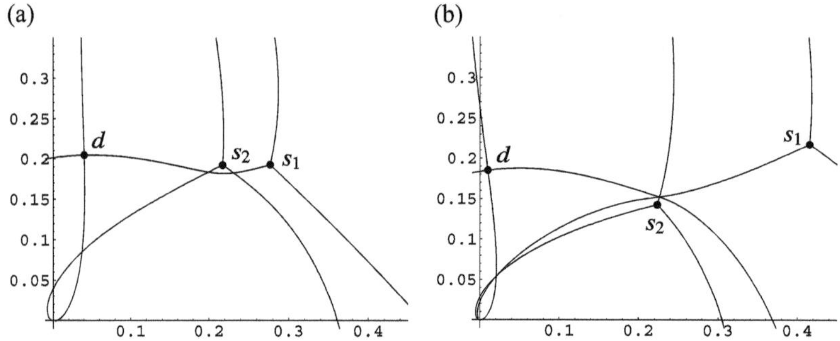

Fig. 2.4 Configuration of Stokes curves of (2.2.2) (**a**) at $t_1 = -1.6104 - 0.2268i$ and (**b**) at $t_3 = -1.5783 - 0.4130i$

let the parameter t vary from t_1 to $t_3 = -1.5783 - 0.4130i$ on the Stokes curve of (2.2.1) in question, we find the turning points d and s_1 are no longer connected by a Stokes curve and instead a trilemma-type crossing point appears at $t = t_3$ (cf. Fig. 2.4b). This trilemma-type crossing point is generated by the "**bifurcation phenomena of Stokes curves**" discussed in [VTP, Section 1.7]. As a matter of fact, during the change from $t = t_1$ to $t = t_3$ a simple turning point s_2 crosses the Stokes curve connecting d and s_1, say at $t = t_2$. If we add relevant virtual turning points and new Stokes curves to Fig. 2.4, at this particular point $t = t_2$ the double turning point d is connected both with s_1 and a virtual turning point v_1. (Similarly, s_1 is connected with d and another virtual turning point v_2; cf. Fig. 2.5b.) The role of an ordinary Stokes curve emanating from d and that of a new Stokes curve emanating from v_2 are interchanged at $t = t_2$ and, when t reaches $t = t_3$, the target of the Stokes curve emanating from d is switched from s_1 to v_1. Consequently, a trilemma-type crossing point appears at $t = t_3$. At a trilemma-type crossing point of Stokes curves, ordinary turning points are connected with virtual turning points and such a trilemma-type crossing can be regarded as a typical degenerate configuration of Stokes curves for a higher-order equation or a system of first-order equations with size greater than or equal to three.

This is an initiating observation of Shunsuke Sasaki. He first discovers trilemma-type crossing and points out the importance of virtual turning points in the theory of degeneration of the Stokes geometry for higher-order equations. Trilemma-type crossing is now understood to be the simplest example of (bidirectional) binary trees. It is generated from a Stokes segment connecting two turning points of the same type via the bifurcation phenomena of Stokes curves. A bifurcation phenomenon of Stokes curves also occurs with trilemma-type crossing and a new binary tree of higher degree is generated as its consequence. This is a typical phenomenon of "**growing trees**" discussed in [VTP, Chapter 2]. Thus the theory of growing trees is considered to be the theory of degeneration of the Stokes geometry for higher-order equations.

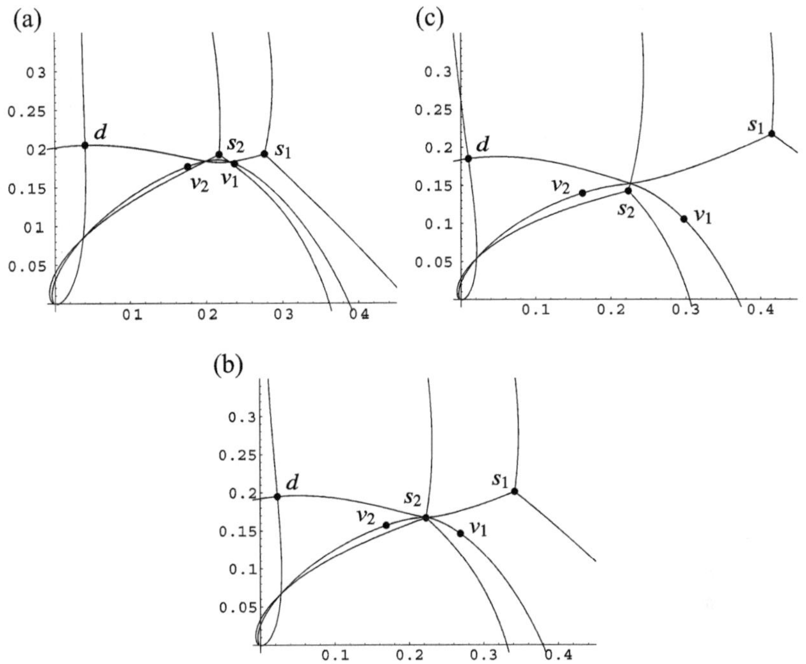

Fig. 2.5 Stokes geometry of (2.2.2) with relevant virtual turning points added (Figure (**a**) is for $t = t_1$, (**b**) for $t = t_2$ and (**c**) for $t = t_3$.)

2.3 Degeneration of the Stokes Geometry and its Relation to the Location of Singularities of the Relevant Borel Transformed WKB Solutions

As was explained in the preceding section, a bidirectional binary tree describes the degeneration of the Stokes geometry for higher-order equations and a trilemma-type crossing is the simplest example of them. Similarly to the case of second-order equations, such a degenerate configuration of Stokes curves for higher-order equations is related to the reality of a period and also to the location of singularities of Borel transformed WKB solutions. By taking the $(1, 4)$ hypergeometric system $\mathcal{M}_{(1,4)HG}$ discussed in §1.5.4 as a particular example, we explain this intriguing relation between the degenerate Stokes geometry, periods, and singularities of Borel transformed WKB solutions in this section.

Let

$$\left(3\frac{d^3}{dx^3} + 2c\eta\frac{d^2}{dx^2} + x\eta^2\frac{d}{dx} - \alpha\eta^3\right)\psi = 0 \tag{2.3.1}$$

be the first equation of the (1, 4) hypergeometric system $\mathcal{M}_{(1,4)HG}$ given by (1.5.5), that is, an equation equivalent to the tangential system $\mathcal{N}_{(1,4)HG}[x_2 = c]$ of $\mathcal{M}_{(1,4)HG}$ to the hyperplane $\{(x_1, x_2) \in \mathbb{C}^2 \; ; \; x_2 = c\}$. As was observed in Sect. 1.5.4, Eq. (2.3.1) has a trilemma-type crossing point P of three Stokes curves when the parameter α is purely imaginary, say when $\alpha = -i$. To be more specific, P is a crossing point of three Stokes curves of type $2 < 3$, $3 < 1$ and $1 < 2$ that emanate from three simple turning points $\hat{x}_1$, $\hat{x}_2$ and $\hat{x}_3$ of type $2 = 3$, $3 = 1$ and $1 = 2$, respectively, as shown in Fig. 2.6. Then, letting $\zeta_j(x)$ $(j = 1, 2, 3)$ denote characteristic roots of

$$p(x, \zeta) = 3\zeta^3 + 2c\zeta^2 + x\zeta - \alpha = 0, \tag{2.3.2}$$

we have

$$\int_{\hat{x}_1}^{P} (\zeta_3(x) - \zeta_2(x))dx, \quad \int_{\hat{x}_2}^{P} (\zeta_1(x) - \zeta_3(x))dx, \quad \int_{\hat{x}_3}^{P} (\zeta_2(x) - \zeta_1(x))dx > 0.$$

Using these integrals, we define

$$\varpi = \int_{\hat{x}_1}^{P} (\zeta_3(x) - \zeta_2(x))dx + \int_{\hat{x}_2}^{P} (\zeta_1(x) - \zeta_3(x))dx + \int_{\hat{x}_3}^{P} (\zeta_2(x) - \zeta_1(x))dx > 0. \tag{2.3.3}$$

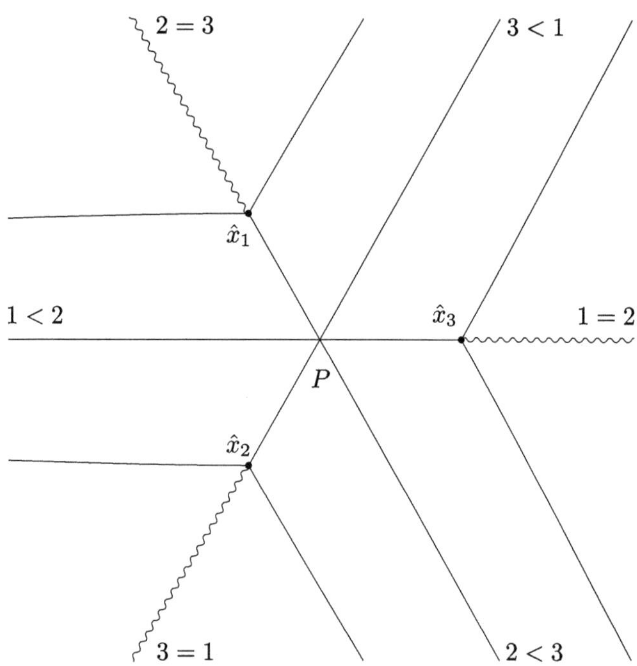

Fig. 2.6 Trilemma-type crossing point of $\mathcal{N}_{(1,4)HG}[x_2 = c]$ when $\alpha = -i$ and $c = 0$ (Wiggly lines designate cuts to define characteristic roots of (2.3.2) as single-valued functions.)

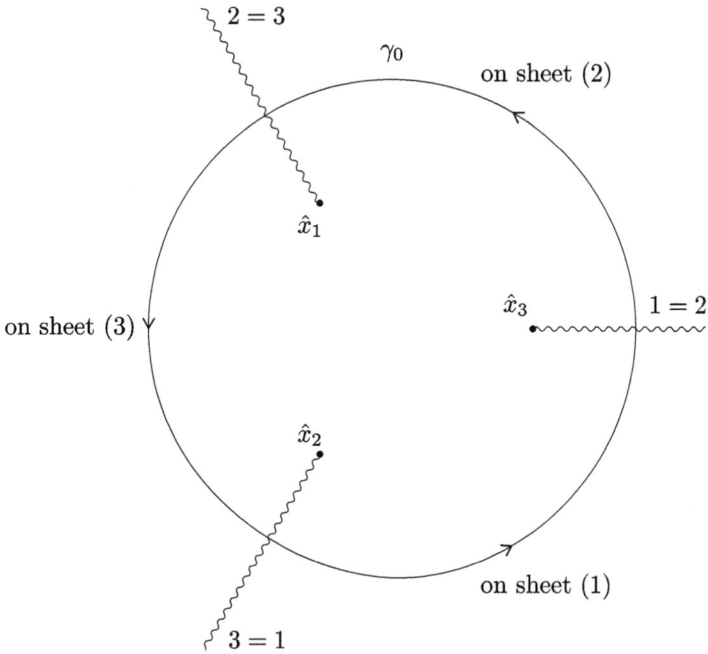

Fig. 2.7 Contour γ_0

This quantity ϖ gives the period of (2.3.1) or the $(1, 4)$ hypergeometric system $\mathcal{M}_{(1,4)HG}$. In fact, ϖ coincides with a contour integral

$$\varpi = \int_{\gamma_0} \zeta(x)\,dx \tag{2.3.4}$$

of the characteristic root $\zeta(x)$ of (2.3.2) along a contour γ_0 on the Riemann surface of $\zeta(x)$ (cf. Fig. 2.7). Furthermore, a trilemma-type crossing point P appears when ϖ is real. Otherwise stated, a trilemma-type crossing point of (2.3.1) is controlled by the reality of this period ϖ.

Remark 2.1 In the case where a bidirectional binary tree T of higher degree appears instead of a trilemma-type crossing, the period is given by its **total integral value** $\Phi(T)$ defined in [VTP, Definition 2.4.4].

Let us compute the explicit value of the period ϖ. We first note that after the analytic continuation from $\hat{x}_1$ to $\hat{x}_2$ along the right-hand portion of γ_0, that is, along a path from $\hat{x}_1$ to $\hat{x}_2$ passing through the right-hand side of $\hat{x}_3$, not only the third branch $\zeta_3(x)$ but also the second branch $\zeta_2(x)$ of the characteristic root of (2.3.2) are ramified at a turning point $\hat{x}_2$. This means that we can replace the cut plane in Fig. 2.7 for the characteristic root $\zeta(x)$ of (2.3.2) by another one shown in Fig. 2.8. This also implies that the Riemann surface of $\zeta(x)$ has two points corresponding

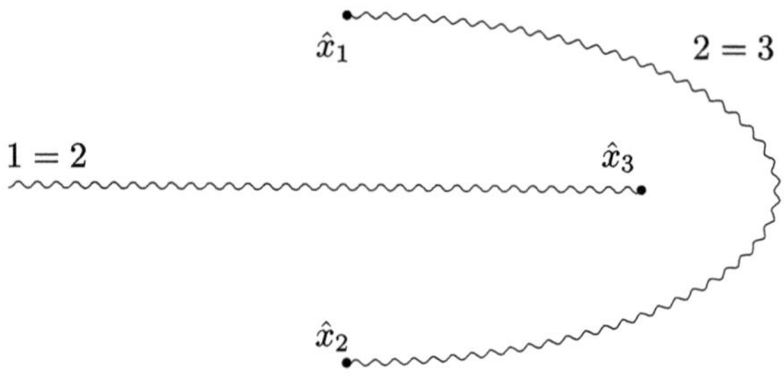

Fig. 2.8 New cut plane for the characteristic root $\zeta(x)$ of (2.3.2)

to ∞: One is a ramification point $x = \infty_1$ for $\zeta_1(x)$ and $\zeta_2(x)$, and another is a non-ramification point $x = \infty_2$ on the third sheet $\zeta_3(x)$ (according to the new cut plane introduced in Fig. 2.8). As a matter of fact, in view of (2.3.2), we find there exist the following two local expressions of $\zeta(x)$ around $x = \infty$:

$$\zeta(x) = \begin{cases} \pm\dfrac{1}{\sqrt{3}i}x^{1/2}\big(1 + O(x^{-1/2})\big), \\ \dfrac{\alpha}{x}\big(1 + O(x^{-1})\big). \end{cases}$$

The first expression corresponds to $\zeta_1(x)$ and $\zeta_2(x)$, while the second expression corresponds to $\zeta_3(x)$. In particular, we have

$$\operatorname*{Res}_{x=\infty_2} \zeta_3(x)\,dx = \operatorname*{Res}_{w=0} \alpha w\big(1 + O(w)\big)\Big(-\frac{1}{w^2}\Big)dw = -\alpha.$$

Since the contour integral (2.3.4) is expressed as an integral of the branch $\zeta_3(x)$ along a contour around $x = \infty_2$, that is, a small clockwise contour around $w = 0$, we obtain

$$\begin{aligned} \varpi &= \int_{\gamma_0} \zeta(x)\,dx = \int_{\text{around } x=\infty_2} \zeta_3(x)\,dx \\ &= -\int_{\text{around } w=0} \zeta_3\Big(\frac{1}{w}\Big)\Big(-\frac{1}{w^2}\Big)dw = 2\pi i\alpha. \end{aligned} \tag{2.3.5}$$

Thus Eq. (2.3.1) has the periods $(2\pi i\alpha)\mathbb{Z}$, i.e., integral multiples of $2\pi i\alpha$.

This period $\varpi = 2\pi i\alpha$ is related also to the location of singularities of Borel transformed WKB solutions of (2.3.1). To see this, we again investigate the bicharacteristics of (2.3.1). For the sake of later convenience, we here compute the bicharacteristic manifold of the $(1, 4)$ hypergeometric system $\mathscr{M}_{(1,4)HG}$. In what follows let us write $\alpha = 3\beta^3$. Then the bicharacteristic manifold of $\mathscr{M}_{(1,4)HG}$ (i.e., of (1.5.5)) is, by definition, a solution manifold of

$$\begin{aligned}&\frac{\partial x_1}{\partial t_j}=\frac{\partial \sigma_j}{\partial \xi_1},\ \frac{\partial x_2}{\partial t_j}=\frac{\partial \sigma_j}{\partial \xi_2},\ \frac{\partial y}{\partial t_j}=\frac{\partial \sigma_j}{\partial \eta},\\ &\frac{\partial \xi_1}{\partial t_j}=-\frac{\partial \sigma_j}{\partial x_1},\ \frac{\partial \xi_2}{\partial t_j}=-\frac{\partial \sigma_j}{\partial x_2},\ \frac{\partial \eta}{\partial t_j}=-\frac{\partial \sigma_j}{\partial y}\end{aligned}\qquad (j=1,2) \tag{2.3.6}$$

that is contained in

$$V=\{(x_1,x_2,y;\xi_1,\xi_2,\eta)\in T^*\mathbb{C}^3\ ;\ \sigma_j(x_1,x_2,\xi_1,\xi_2,\eta)=0\ (j=1,2),\ \eta\neq 0\} \tag{2.3.7}$$

and satisfies an initial condition, say

$$(x_1,x_2,y;\xi_1,\xi_2,\eta)\Big|_{t_1=t_2=0}=(0,0,0;\beta,\beta^2,1), \tag{2.3.8}$$

where

$$\sigma_1(x_1,x_2,\xi_1,\xi_2,\eta)=3\xi_1^3+2x_2\xi_1^2+x_1\xi_1\eta^2-3\beta^3\eta^3, \tag{2.3.9}$$

$$\sigma_2(x_1,x_2,\xi_1,\xi_2,\eta)=\xi_2\eta-\xi_1^2 \tag{2.3.10}$$

denote the principal symbols of the Borel transform of (1.5.5). We first solve

$$\begin{aligned}&\frac{\partial x_1}{\partial t_2}=\frac{\partial \sigma_2}{\partial \xi_1}=-2\xi_1,\ \frac{\partial x_2}{\partial t_2}=\frac{\partial \sigma_2}{\partial \xi_2}=\eta,\ \frac{\partial y}{\partial t_2}=\frac{\partial \sigma_2}{\partial \eta}=\xi_2,\\ &\frac{\partial \xi_1}{\partial t_2}=-\frac{\partial \sigma_2}{\partial x_1}=0,\ \frac{\partial \xi_2}{\partial t_2}=-\frac{\partial \sigma_2}{\partial x_2}=0,\ \frac{\partial \eta}{\partial t_2}=-\frac{\partial \sigma_2}{\partial y}=0\end{aligned}$$

with the initial condition (2.3.8) to obtain

$$x_1=-2\beta t_2,\ \ x_2=t_2,\ \ y=\beta^2 t_2,\ \ \xi_1=\beta,\ \ \xi_2=\beta^2,\ \ \eta=1. \tag{2.3.11}$$

Then we solve

$$\begin{aligned}\frac{\partial x_1}{\partial t_1}&=\frac{\partial \sigma_1}{\partial \xi_1}=9\xi_1^2+4x_2\xi_1\eta+x_1\eta^2,\\ \frac{\partial x_2}{\partial t_1}&=\frac{\partial \sigma_1}{\partial \xi_2}=0,\\ \frac{\partial y}{\partial t_1}&=\frac{\partial \sigma_1}{\partial \eta}=2x_2\xi_1^2+2x_1\xi_1\eta-9\beta^3\eta^2,\\ \frac{\partial \xi_1}{\partial t_1}&=-\frac{\partial \sigma_1}{\partial x_1}=-\xi_1\eta^2,\end{aligned}$$

$$\frac{\partial \xi_2}{\partial t_1} = -\frac{\partial \sigma_1}{\partial x_2} = -2\xi_1^2\eta,$$

$$\frac{\partial \eta}{\partial t_1} = -\frac{\partial \sigma_1}{\partial y} = 0,$$

regarding (2.3.11) as its initial condition. By the second and the last equations, we readily find $x_2 = t_2$ and $\eta = 1$. Then the fourth and the fifth equations imply $\xi_1 = \beta e^{-t_1}$ and $\xi_2 = \beta^2 e^{-2t_1}$. Finally, it follows from the first and the third equations that

$$x_1 = -3\beta^2 e^{-2t_1} - 2\beta e^{-t_1} t_2 + 3\beta^2 e^{t_1},$$

$$y = 2\beta^3 e^{-3t_1} + \beta^2 e^{-2t_1} t_2 - 3\beta^3 t_1 - 2\beta^3.$$

Hence the bicharacteristic manifold of $\mathscr{M}_{(1,4)HG}$ is described as

$$\begin{aligned} &x_1 = -3\beta^2 e^{-2t_1} - 2\beta e^{-t_1} t_2 + 3\beta^2 e^{t_1}, \quad x_2 = t_2, \\ &y = 2\beta^3 e^{-3t_1} + \beta^2 e^{-2t_1} t_2 - 3\beta^3 t_1 - 2\beta^3, \\ &\xi_1 = \beta e^{-t_1}, \quad \xi_2 = \beta^2 e^{-2t_1}, \quad \eta = 1. \end{aligned} \tag{2.3.12}$$

Putting $x_2 = t_2 = c$ into (2.3.12), we thus obtain the bicharacteristic strip of the Borel transform of (2.3.1):

$$\begin{aligned} &x = -3\beta^2 e^{-2t} - 2\beta c e^{-t} + 3\beta^2 e^{t}, \quad y = 2\beta^3 e^{-3t} + \beta^2 c e^{-2t} - 3\beta^3 t - 2\beta^3, \\ &\xi = \beta e^{-t}, \quad \eta = 1. \end{aligned} \tag{2.3.13}$$

In particular,

$$x = -3\beta^2 e^{-2t} - 2\beta c e^{-t} + 3\beta^2 e^{t}, \quad y = 2\beta^3 e^{-3t} + \beta^2 c e^{-2t} - 3\beta^3 t - 2\beta^3 \tag{2.3.14}$$

describes the bicharacteristic curve of the Borel transform of (2.3.1). If we solve the first equation of (2.3.14) with respect to t for each fixed x, we get three solutions for e^{-t}:

$$e^{-t} = \phi_j(x;\beta,c) \quad (j = 1, 2, 3).$$

Then, substitution of this solution to the second equation of (2.3.14) gives

$$y_j(x) = 2\beta^3 \phi_j(x;\beta,c)^3 + \beta^2 c \phi_j(x;\beta,c)^2 + 3\beta^3 \log \phi_j(x;\beta,c) - 2\beta^3. \tag{2.3.15}$$

This formula expresses the singularities $y = y_j(x)$ $(j = 1, 2, 3)$ of Borel transformed WKB solutions of (2.3.1) in the Borel plane. In view of the multivaluedness of the logarithmic function, we find that the singularities of Borel transformed WKB solutions have a periodic structure with the period $(3\beta^3)2\pi i\mathbb{Z} = 2\pi i\alpha\mathbb{Z}$. Thus the period $\varpi = 2\pi i\alpha$ appears in the description of the location of singularities of Borel transformed WKB solutions of (2.3.1). When the period $\varpi = 2\pi i\alpha$ is real, the Stokes geometry of (2.3.1) becomes degenerate, that is, we have a trilemma-type crossing point, and consequently the Borel summability of WKB solutions breaks down.

Remark 2.2 When there is a trilemma-type crossing point, the Borel summability of WKB solutions breaks down and a **parametric Stokes phenomenon** occurs with WKB solutions. The explicit form of the connection formula describing a parametric Stokes phenomenon for a trilemma-type crossing is provided in the paper [Sa1].

Having this expression of singularities of Borel transformed WKB solutions in terms of the bicharacteristic curve in mind, we obtain Fig. 2.9, which schematically describes the location of singularities of Borel transformed WKB solutions around a trilemma-type crossing point in $\mathbb{C}^2_{(x,y)}$. At each simple turning point $\hat{x}_j$ two branches of the bicharacteristic curve (i.e., the singularity of Borel transformed WKB solutions) meet and have the shape of a cusp, while at a trilemma-type crossing point P each branch of the bicharacteristic curve coming from simple

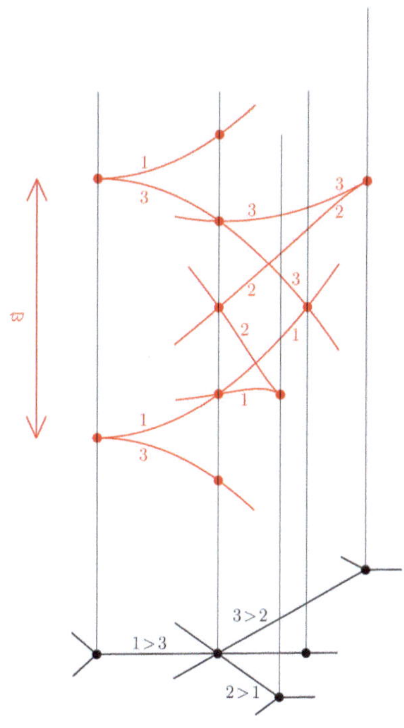

Fig. 2.9 Schematic picture for singularities of Borel transformed WKB solutions around a trilemma-type crossing point (The horizontal plane is the x-plane and the vertical line is the y-plane.)

turning points is connected smoothly. Now, let us start from one simple turning point, say $\hat{x}_2$, and make a round trip

$$\hat{x}_2 \to P \to \hat{x}_3 \to P \to \hat{x}_1 \to P \to \hat{x}_2 \tag{2.3.16}$$

along the relevant branches of the bicharacteristic curve. Then, when we return to the starting point $\hat{x}_2$, the position of y is shifted by ϖ, which is nothing but the period defined by the integral relation (2.3.3). Figure 2.9 visualizes this situation clearly and explains why the period ϖ appears in the location of singularities of Borel transformed WKB solutions.

In contrast with this, the schematic picture of the bicharacteristic curve around a triordered-type crossing point is quite different from that around a trilemma-type crossing point. See Fig. 2.10, which shows the schematic picture around a triordered-type crossing point. In the case of a triordered-type crossing point, if we make the same round trip as (2.3.16) along the relevant branches of the bicharacteristic curve, the position of y is not shifted when we return to the starting point $\hat{x}_2$. Otherwise stated, no period appears with singularities of Borel transformed WKB solutions around a triordered-type crossing point. This means that a triordered-type crossing point is not related to degeneration of the Stokes geometry.

Fig. 2.10 Schematic picture for singularities of Borel transformed WKB solutions around a triordered-type crossing point

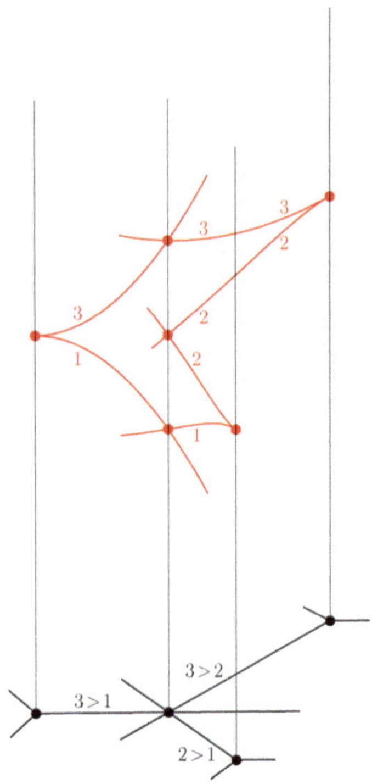

Chapter 3
A Bicharacteristic Chain Associated with a Non-hereditary Turning Point

Abstract In Chap. 1 we considered the tangential systems $\mathcal{N}(c)$ and $\mathcal{N}_{(1,4)HG}[x_1 = c]$ of the Pearcey system and the (1,4) hypergeometric system, respectively, and observed that these tangential systems have periods. Then in Chap. 2 we verified that these periods are defined through the bicharacteristics of their Borel transforms and also expressed as contour integrals on the Riemann surface of the characteristic root. In this chapter we discuss the relationship of these periods with the Stokes geometry and the periodic structure for singularities of solutions of their Borel transforms in a more detailed manner. First, for $\mathcal{N}(c)$ we introduce its perturbed system $\widetilde{\mathcal{N}(c)}$ and consider its bicharacteristic chain to show that there exist additional periods and virtual turning points for $\widetilde{\mathcal{N}(c)}$ which are associated with a non-hereditary turning point of $\mathcal{N}(c)$ and do not originate from the virtual turning points of the original Pearcey system. We also discuss the relationship between such additional virtual turning points and the Stokes geometry. Then, for $\mathcal{N}_{(1,4)HG}[x_1 = c]$ we show that its perturbed system also has additional periods similarly to $\mathcal{N}(c)$ and discuss their relationship with the degeneration of the Stokes geometry. Finally, in Sect. 3.4 we derive connection formulas related to the leaf-type and tadpole-type configurations that appeared in Sect. 1.5.

3.1 A Period and a New Class of Virtual Turning Points Associated with a Non-hereditary Turning Point

In this section, by considering a perturbed system $\widetilde{\mathcal{N}(c)}$ of the tangential system $\mathcal{N}(c)$ of the Pearcey system, we discuss periods and virtual turning points associated with a non-hereditary turning point of $\mathcal{N}(c)$.

First, let us recall the explicit form of $\mathcal{N}(c)$. The tangential system $\mathcal{N}(c)$ is given by

$$P^{(c)}\psi = 0$$

S. Hirose et al., *Virtual Turning Points II*, SpringerBriefs in Mathematical Physics 52, https://doi.org/10.1007/978-981-95-2817-2_3

with

$$
\begin{aligned}
P^{(c)} =&16c^3 \left(\eta^{-1}\frac{d}{dz}\right)^3 + 16c^5(z-1)\left(\eta^{-1}\frac{d}{dz}\right)^2 \\
&+ 4c^4\left\{c^3z^2 - (2c^3-5)z + c^3 - 2\right\}\eta^{-1}\frac{d}{dz} + c^3\left\{c^3z^2 - 2(c^3-2)z\right\} + Q,
\end{aligned}
$$

where $Q = \sum_{j\geq 1}\eta^{-j}q_j(z, \eta^{-1}d/dz)$ consists of **WKB-theoretically lower-order terms**. In what follows the **WKB-theoretically principal part** of $P^{(c)}$ is denoted by $P_0^{(c)}$. For a generic value of c the tangential system $\mathcal{N}(c)$ has three simple turning points which are zeros of

$$
8c^3z^3 + (27 - 24c^3)z^2 + 24c^3z - 8c^3
$$

and a double turning point $z = (2 - 4c^{-3})/3$. This double turning point is a non-hereditary turning point. As was discussed in Sect. 1.2, no Stokes phenomenon occurs on a Stokes curve emanating from a non-hereditary turning point. However, it is shown in [HiKST] that Stokes curves emanating from the non-hereditary turning point may become relevant to Stokes phenomena when WKB-theoretically lower-order terms are added to $\mathcal{N}(c)$. We denote by $\widetilde{\mathcal{N}(c)}$ the perturbed system of $P^{(c)}\psi = 0$ obtained by adding WKB-theoretically lower-order terms. In discussing the perturbed system $\widetilde{\mathcal{N}(c)}$, we encounter ordered crossing points arising from the non-hereditary turning point and consequently much more virtual turning points and new Stokes curves appear with $\widetilde{\mathcal{N}(c)}$. We also note that the Borel transform of $\widetilde{\mathcal{N}(c)}$ is with double characteristics (cf. Sect. 1.4). Hence singularities of solutions of the Borel transform of $\widetilde{\mathcal{N}(c)}$ propagate along a **bicharacteristic chain** rather than a bicharacteristic curve, as is shown in [KKO]. This means that it is necessary to use a bicharacteristic chain instead of a bicharacteristic curve to define a virtual turning point of $\widetilde{\mathcal{N}(c)}$. As the computation of virtual turning points of $\widetilde{\mathcal{N}(c)}$ in terms of the bicharacteristic chain is already discussed in [HiKST], let us here recall the argument in [HiKST].

As discussed in Sect. 1.4, the principal symbol $p_0^{(c)}(z, \zeta, \eta)$ of the Borel transform of $P_0^{(c)}$ has the factorization

$$
p_0^{(c)}(z, \zeta, \eta) = -q_+(z, \zeta, \eta)q_-(z, \zeta, \eta),
$$

where

$$
q_\pm(z, \zeta, \eta) = (6c\eta^{-1}\zeta + 2 - c^3) \pm (2c\eta^{-1}\zeta + 2 - c^3 + c^3z)\sqrt{4c\eta^{-1}\zeta + 1}.
$$

Furthermore, the bicharacteristic strip of $q_\pm(z, \zeta, \eta)$ with the initial condition

$$
(z(0), y(0), \zeta(0), \eta(0)) = \left(\frac{2-4c^{-3}}{3}, 0, \frac{c^2 - 2c^{-1}}{6}, 1\right)
$$

is given by

$$\begin{aligned} \mathscr{b}_\pm : \quad & (z_\pm(\sigma), y_\pm(\sigma), \zeta_\pm(\sigma), \eta_\pm(\sigma)) \\ &= \left(\tilde{z}(\sigma \mp \alpha), \tilde{y}(\sigma \mp \alpha) - \tilde{y}(\mp\alpha), \tilde{\zeta}(\sigma \mp \alpha), 1\right), \end{aligned}$$

where

$$\begin{aligned} \tilde{z}(\tilde{\sigma}) &= -2c^5\tilde{\sigma}^2 + 3c\tilde{\sigma} + \left(1 - \frac{3}{2c^3}\right) + \left(-\frac{1}{2c^4} + \frac{1}{4c^7}\right)\frac{1}{\tilde{\sigma}}, \\ \tilde{y}(\tilde{\sigma}) &= c^{12}\tilde{\sigma}^4 - c^8\tilde{\sigma}^3 - \frac{1}{2}c^4\tilde{\sigma}^2 + \left(1 - \frac{c^3}{2}\right)\tilde{\sigma} + \left(-\frac{1}{8c^5} + \frac{1}{16c^8}\right)\frac{1}{\tilde{\sigma}}, \\ \tilde{\zeta}(\tilde{\sigma}) &= c^7\tilde{\sigma}^2 - \frac{1}{4c}, \\ \alpha &= \frac{1}{2c^4}\left(\frac{2c^3 - 1}{3}\right)^{1/2}. \end{aligned}$$

Note that we can confirm the following relations hold:

$$z_+\Big|_{\sigma=0} = z_-\Big|_{\sigma=0}, \quad y_+\Big|_{\sigma=0} = y_-\Big|_{\sigma=0},$$

$$\frac{dy_+}{d\sigma}\Big/\frac{dz_+}{d\sigma}\Big|_{\sigma=0} = \frac{dy_-}{d\sigma}\Big/\frac{dz_-}{d\sigma}\Big|_{\sigma=0}.$$

Hence the bicharacteristic curves $\pi(\mathscr{b}_+)$ and $\pi(\mathscr{b}_-)$, that is, the projection to $\mathbb{C}^2_{(z,y)}$ of $\mathscr{b}_+$ and $\mathscr{b}_-$, respectively, are mutually tangent at $(z, y) = (z(0), y(0)) = ((2 - 4c^{-3})/3, 0)$. Figure 3.1 shows the configuration of two bicharacteristic curves $\pi(\mathscr{b}_\pm)$ near the point $(z, y) = (\hat{z}, 0) = ((2 - 4c^{-3})/3, 0)$. Hence singularities of solutions of the Borel transform of $\widetilde{\mathscr{N}(c)}$ bifurcate at this point. From the explicit form of $\mathscr{b}_\pm$ we also find that

$$\begin{aligned} &(z_-(\sigma - 2\alpha), y_-(\sigma - 2\alpha), \zeta_-(\sigma - 2\alpha), \eta_-(\sigma - 2\alpha)) \\ &\qquad = (z_+(\sigma), y_+(\sigma) - (\tilde{y}(\alpha) - \tilde{y}(-\alpha)), \zeta_+(\sigma), \eta_+(\sigma)) \end{aligned}$$

holds. This means that, if we let ϖ denote

$$\varpi = \tilde{y}(\alpha) - \tilde{y}(-\alpha),$$

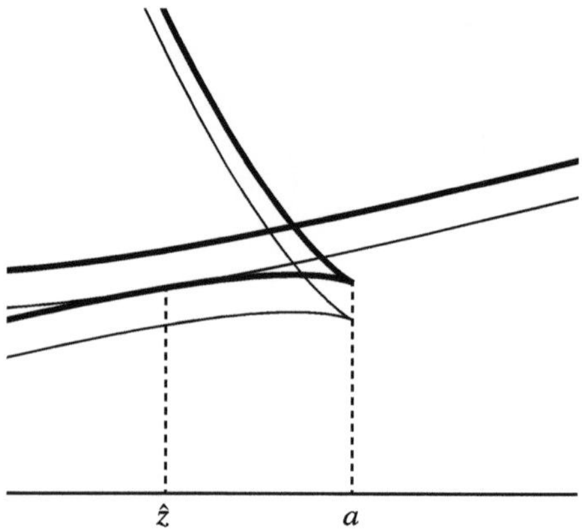

Fig. 3.1 Bicharacteristic curves $\pi(б_+)$ and $\pi(б_-)$, which are designated by thick lines and thin lines, respectively. Here $\hat{z}$ denotes the non-hereditary double turning point and a denotes a simple turning point

then $б_-$ can be considered as the bicharacteristic strip of $q_+(z, \zeta, \eta)$ with the initial condition

$$\left(\frac{2 - 4c^{-3}}{3}, -\varpi, \frac{c^2 - 2c^{-1}}{6}, 1\right)$$

(up to the shift by 2α of the time variable σ). Similarly, $б_+$ can be considered as the bicharacteristic strip of $q_-(z, \zeta, \eta)$ with the initial condition

$$\left(\frac{2 - 4c^{-3}}{3}, \varpi, \frac{c^2 - 2c^{-1}}{6}, 1\right)$$

(up to the shift of the time variable again).

Let us now consider the bicharacteristic strip $б_m : (z_m(\sigma), y_m(\sigma), \zeta_m(\sigma), \eta_m(\sigma))$ of $q_-(z, \zeta, \eta)$ with the initial condition

$$\left(\frac{2 - 4c^{-3}}{3}, m\varpi, \frac{c^2 - 2c^{-1}}{6}, 1\right) \tag{3.1.1}$$

for $m \in \mathbb{Z}$. This is a generalization of $б_\pm$ to the effect that $б_+$ and $б_-$ can be considered as $б_1$ and $б_0$, respectively. The bicharacteristic strip $б_m$ is explicitly given by

$$\begin{aligned}
&(z_m(\sigma), y_m(\sigma), \zeta_m(\sigma), \eta_m(\sigma)) \\
&\quad = (z_-(\sigma), y_-(\sigma) + m\varpi, \zeta_-(\sigma), \eta_-(\sigma)) \\
&\quad = \left(\tilde{z}(\sigma + \alpha), \tilde{y}(\sigma + \alpha) - \tilde{y}(\alpha) + m\varpi, \tilde{\zeta}(\sigma + \alpha), 1\right).
\end{aligned}$$

Since

$$(z_m(\sigma - 2\alpha), y_m(\sigma - 2\alpha), \zeta_m(\sigma - 2\alpha), \eta_m(\sigma - 2\alpha)) \\ = (z_+(\sigma), y_+(\sigma) + (m-1)\varpi, \zeta_+(\sigma), \eta_+(\sigma))$$

holds, $\mathscr{b}_m$ can be considered also as the bicharacteristic strip of $q_+(z, \zeta, \eta)$ with the initial condition

$$\left(\frac{2-4c^{-3}}{3}, (m-1)\varpi, \frac{c^2 - 2c^{-1}}{6}, 1\right).$$

Further, since we have

$$\begin{aligned}&(z_m(0), y_m(0), \zeta_m(0), \eta_m(0)) \\ &\quad = (z_{m+1}(-2\alpha), y_{m+1}(-2\alpha), \zeta_{m+1}(-2\alpha), \eta_{m+1}(-2\alpha)) \\ &\quad = \left(\frac{2-4c^{-3}}{3}, m\varpi, \frac{c^2 - 2c^{-1}}{6}, 1\right)\end{aligned}$$

and

$$\frac{dy_m}{d\sigma}\Big/\frac{dz_m}{d\sigma}\Big|_{\sigma=0} = \frac{dy_-}{d\sigma}\Big/\frac{dz_-}{d\sigma}\Big|_{\sigma=0} = \frac{dy_+}{d\sigma}\Big/\frac{dz_+}{d\sigma}\Big|_{\sigma=0} = \frac{dy_{m+1}}{d\sigma}\Big/\frac{dz_{m+1}}{d\sigma}\Big|_{\sigma=-2\alpha},$$

we find that the bicharacteristic strips $\mathscr{b}_m$ and $\mathscr{b}_{m+1}$ both pass through the point

$$\left(\frac{2-4c^{-3}}{3}, m\varpi, \frac{c^2 - 2c^{-1}}{6}, 1\right)$$

and the bicharacteristic curves $\pi(\mathscr{b}_m)$ and $\pi(\mathscr{b}_{m+1})$ are mutually tangent at $(z, y) = ((2-4c^{-3})/3, m\varpi)$. Hence singularities of solutions of the Borel transform of $\widetilde{\mathcal{N}}(c)$ bifurcate at this point and propagate along all the bicharacteristic curves $\pi(\mathscr{b}_m)$. We thus consider the bicharacteristic chain consisting of all $\pi(\mathscr{b}_m)$:

$$\bigcup_{m\in\mathbb{Z}} \pi(\mathscr{b}_m)$$

or, more precisely, the collection of $\pi(\mathscr{b}_m)$ with the parametrization

$$\pi(\mathscr{b}_m): \quad (z_m(\sigma), y_m(\sigma)) = (\tilde{z}(\sigma+\alpha), \tilde{y}(\sigma+\alpha) - \tilde{y}(\alpha) + m\varpi) \quad (m \in \mathbb{Z}).$$

This representation of the bicharacteristic chain manifestly explains that the singularities of the Borel transform of $\widetilde{\mathcal{N}(c)}$ has a periodic structure described by ϖ. We now claim that ϖ is expressed by the following integral:

$$\varpi = \int_\gamma \zeta(z)dz, \tag{3.1.2}$$

where γ denotes a closed curve on the Riemann surface of the root $\zeta(z)$ of $p_0^{(c)}(z, \zeta, 1) = 0$.

To prove this, we consider the following functions of σ defined by the bicharacteristic strip $\mathscr{b}_-$:

$$z_-(\sigma) = \tilde{z}(\sigma + \alpha), \quad \zeta_-(\sigma) = \tilde{\zeta}(\sigma + \alpha).$$

Using these functions, we find the points $\sigma = 0, -2\alpha$ correspond to the non-hereditary double turning point of $\widetilde{\mathcal{N}(c)}$ since

$$(z_-(\sigma), \zeta_-(\sigma)) = \left(\frac{2 - 4c^{-3}}{3}, \frac{c^2 - 2c^{-1}}{6}\right)$$

holds at $\sigma = 0$ and $\sigma = -2\alpha$. Note that, although there are three values of σ satisfying $z_-(\sigma) = (2 - 4c^{-3})/3$, only two of them, that is, $\sigma = 0$ and $\sigma = -2\alpha$ satisfy $\zeta_-(\sigma) = (c^2 - 2c^{-1})/6$. Thus only $\sigma = 0$ and $\sigma = -2\alpha$ correspond to the non-hereditary turning point. Additionally, if we let $\sigma^{(j)}$ $(j = 1, 2, 3)$ denote a solution of $(dz_-/d\sigma)(\sigma) = 0$, $\sigma^{(j)}$ $(j = 1, 2, 3)$ correspond to simple turning points. Furthermore, $z_-(\sigma)$ also has a pole at $\sigma = -\alpha$. Taking these situations into account, we choose a curve $\sigma = \sigma(t)$ $(0 \le t \le 1)$ in the σ-plane so that $\sigma(t)$ satisfies $\sigma(0) = 0, \sigma(1) = -2\alpha$, and

$$\sigma(t) \in \mathbb{C} \setminus \left\{-\alpha, \sigma^{(1)}, \sigma^{(2)}, \sigma^{(3)}\right\} \quad (0 < t < 1),$$

and consider a curve γ defined by $z = z_-(\sigma(t))$ $(0 \le t \le 1)$. The starting point and endpoint of γ are both the non-hereditary double turning point $(2 - 4c^{-3})/3$. Further, as $(z_-(\sigma), \zeta_-(\sigma))$ is defined by a bicharacteristic strip, on the curve γ an appropriate branch $\zeta = \zeta(z)$ of the root of $p_0^{(c)}(z, \zeta, 1) = 0$ with respect to ζ is chosen so that $\zeta_-(\sigma) = \zeta(z_-(\sigma))$ holds. Therefore, since it follows from the definition of the bicharacteristic strip and the homogeneity of $q_-(z, \zeta, \eta)$ with respect to (ζ, η) that

$$\begin{aligned}
&\zeta_-(\sigma)\frac{dz_-}{d\sigma}(\sigma) + \frac{dy_-}{d\sigma}(\sigma) \\
&= \left.\left(\zeta_-(\sigma)\frac{\partial q_-}{\partial \zeta}(z_-(\sigma), \zeta_-(\sigma), \eta) + \eta\frac{\partial q_-}{\partial \eta}(z_-(\sigma), \zeta_-(\sigma), \eta)\right)\right|_{\eta=1} = 0,
\end{aligned}$$

we obtain

$$\begin{aligned}\int_\gamma \zeta(z)dz &= \int_0^1 \zeta(z_-(\sigma(t)))\frac{d}{dt}z_-(\sigma(t))dt \\ &= \int_0^1 \zeta_-(\sigma(t))\frac{dz_-}{d\sigma}(\sigma(t))\frac{d\sigma}{dt}(t)dt \\ &= -\int_0^1 \frac{dy_-}{d\sigma}(\sigma(t))\frac{d\sigma}{dt}(t)dt \\ &= -\int_0^1 \frac{d}{dt}y_-(\sigma(t))dt \\ &= -\, y_-(\sigma(1)) + y_-(\sigma(0)) \\ &= -\, y_-(-2\alpha) + y_-(0) \\ &= -\, \tilde{y}(-\alpha) + \tilde{y}(\alpha) \\ &= \varpi .\end{aligned}$$

For example, let us consider $\widetilde{\mathcal{N}(c)}$ for $c = \exp\{(22/32)\pi i\}$. In this case there are essentially two choices $\sigma_1(t)$ and $\sigma_2(t)$ for $\sigma(t)$ satisfying the above conditions, as is shown in Fig. 3.2. The corresponding curves γ_1 and γ_2 (precisely speaking, curves path-homotopic to $z_-(\sigma_1(t))$ and $z_-(\sigma_2(t))$, respectively) are depicted in Fig. 3.3. For these curves γ_1 and γ_2 we have

$$\varpi = \int_{\gamma_1} \zeta(z)dz = \int_{\gamma_2} \zeta(z)dz. \tag{3.1.3}$$

The above argument guarantees that the value of the integral is independent of the choice of $\sigma(t)$ as long as all the above conditions are satisfied.

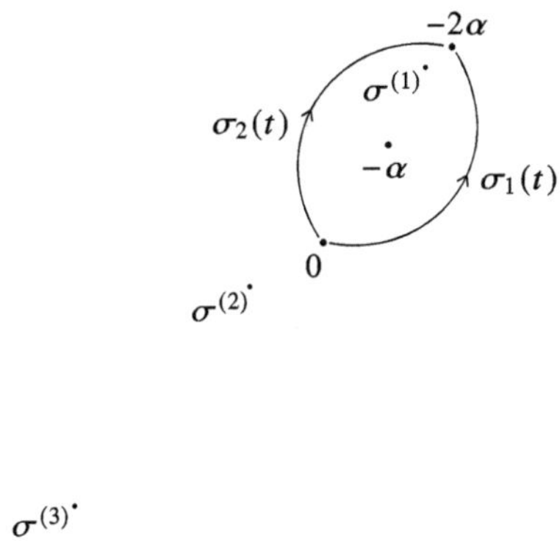

Fig. 3.2 Curves $\sigma_1(t)$ and $\sigma_2(t)$. Here small dots are the solutions $\sigma^{(j)}$ $(j = 1, 2, 3)$ of $(dz_-/d\sigma)(\sigma) = 0$

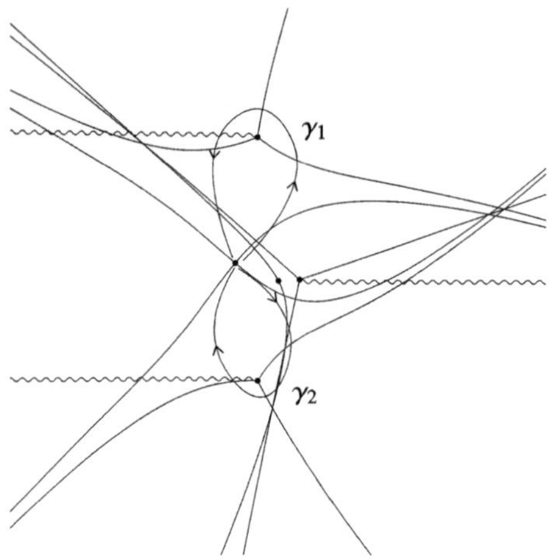

Fig. 3.3 The Stokes geometry of $\widetilde{\mathcal{N}}(c)$ for $c = \exp\{(22/32)\pi i\}$ and the curves γ_1 and γ_2

Note that the Riemann surface of the characteristic root $\zeta(z)$ of $p_0^{(c)}(z, \zeta, 1) = 0$ is pinched at the non-hereditary double turning point $z = (2 - 4c^{-3})/3$. This means that it has a singularity at

$$(z, \zeta) = \left(\frac{2 - 4c^{-3}}{3}, \frac{c^2 - 2c^{-1}}{6}\right).$$

Thus, precisely speaking, $p_0^{(c)}(z, \zeta, 1) = 0$ defines merely an algebraic curve $\mathcal{R}$ with singularity. The curve $(z(t), \zeta(t)) = (z_-(\sigma(t)), \zeta_-(\sigma(t)))$ $(0 \le t \le 1)$ introduced above defines a curve on $\mathcal{R}$ with endpoints at the singularity.

In conclusion, in the case of the perturbed system $\widetilde{\mathcal{N}}(c)$, since the singularities of solutions of its Borel transform propagate along the bicharacteristic chain (cf. [KKO]), it is necessary to use the bicharacteristic chain defined above instead of the bicharacteristic curve in defining its virtual turning points. This leads to the following definition of a virtual turning point of $\widetilde{\mathcal{N}}(c)$: a point z_0 is a **virtual turning point** of $\widetilde{\mathcal{N}}(c)$ if there exist $\sigma_1, \sigma_2 \in \mathbb{C}$ and $m_1, m_2 \in \mathbb{Z}$ $((\sigma_1, m_1) \neq (\sigma_2, m_2))$ such that the following relations hold:

$$z_0 = z_{m_1}(\sigma_1) = z_{m_2}(\sigma_2), \quad y_{m_1}(\sigma_1) = y_{m_2}(\sigma_2).$$

Since $z_m(\sigma) = z_-(\sigma)$ and $y_m(\sigma) = y_-(\sigma) + m\varpi$, these relations can be rewritten as

$$z_0 = z_-(\sigma_1) = z_-(\sigma_2), \quad y_-(\sigma_1) = y_-(\sigma_2) + (m_2 - m_1)\varpi. \tag{3.1.4}$$

It is clear that, without loss of generality, we may assume $m = m_2 - m_1 \geq 0$.

3.2 Stokes Geometry of $\widetilde{\mathcal{N}(c)}$, the Perturbed System of $\mathcal{N}(c)$

In this section, based on the (modified) definition of virtual turning points given at the end of the previous section, we consider some concrete examples of the Stokes geometry of $\widetilde{\mathcal{N}(c)}$ and see that the period discussed in the previous section has close relationship with degeneration of the Stokes geometry.

First, we present the Stokes geometry of $\widetilde{\mathcal{N}(c)}$ for $c = \exp\{(22/32)\pi i\}$ with taking only the virtual turning points of the original Pearcey system, that is, the virtual turning points obtained by setting $m = 0$ in (3.1.4), into account. The result is shown in Fig. 3.4. Note that, if we assume no Stokes phenomenon occurs on Stokes curves emanating from the non-hereditary double turning point, then no ordered crossing point appears in Fig. 3.4.

Next, for a fixed positive integer M, we add the virtual turning points obtained by setting $m = 1, 2, \ldots, M$ in (3.1.4) and the (new) Stokes curves emanating from them to Fig. 3.4. Figures 3.5, 3.6 and 3.7 respectively show the Stokes geometries thus obtained for $M = 1, 2, 3$. The number of virtual turning points appearing in these Stokes geometries is 10 for Fig. 3.5, 21 for Fig. 3.6, and 32 for Fig. 3.7.

So far, we have not considered the activeness of Stokes curves, that is, we have not checked yet whether a Stokes phenomenon occurs on each Stokes curve. We now consider the activeness of Stokes curves and represent the portion of a Stokes curve where Stokes phenomena may occur with a solid line and the portion where Stokes phenomena may not occur with a dotted line. Applying the general rule explained in [VTP] and reviewed in Sect. 1.1, we determine whether each portion of a Stokes curve is represented by a solid line or a dotted line. To be more specific, we apply the following rule:

- A Stokes curve emanating from an ordinary turning point is a solid line near that turning point.

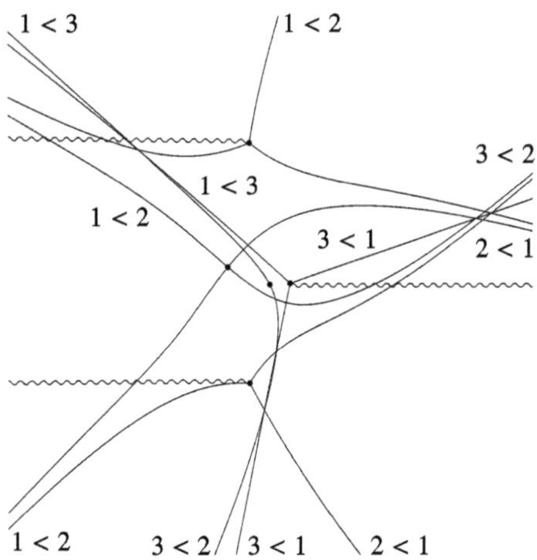

Fig. 3.4 The Stokes geometry of $\widetilde{\mathcal{N}(c)}$ for $c = \exp\{(22/32)\pi i\}$ with the virtual turning points obtained by setting $m = 0$

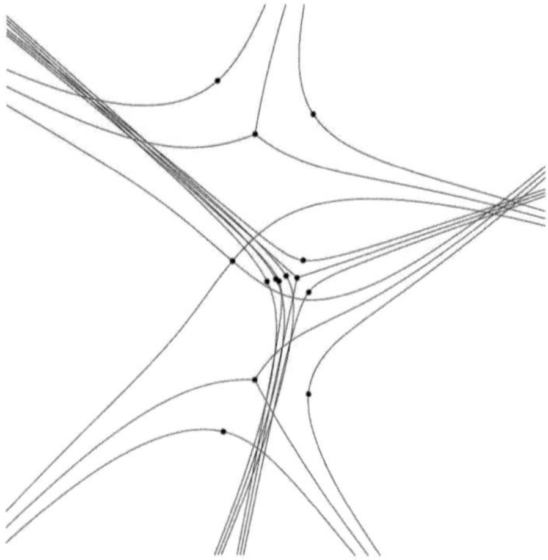

Fig. 3.5 The Stokes geometry with the virtual turning points obtained by setting $m = 0, 1$

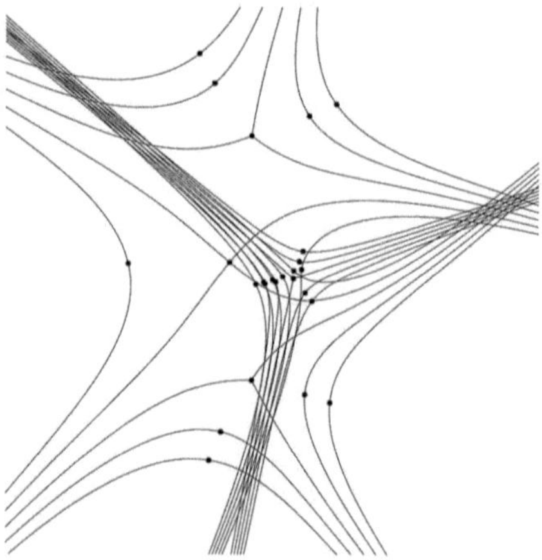

Fig. 3.6 The Stokes geometry with the virtual turning points obtained by setting $m = 0, 1, 2$

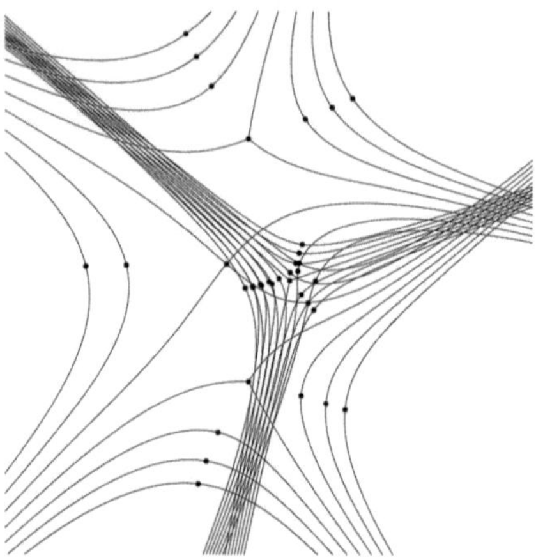

Fig. 3.7 The Stokes geometry with the virtual turning points obtained by setting $m = 0, 1, 2, 3$

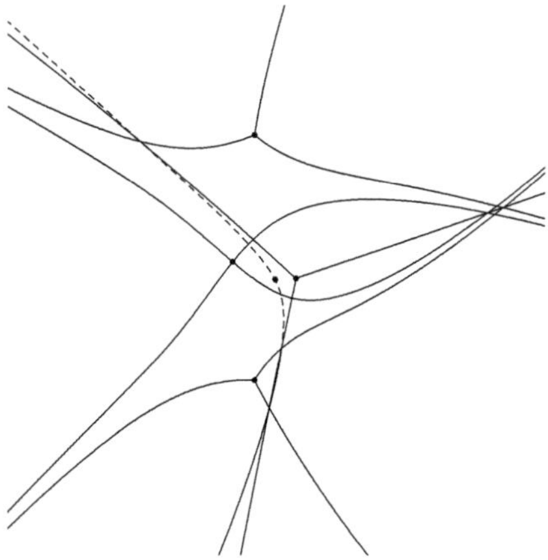

Fig. 3.8 The Stokes geometry obtained from Fig. 3.4 by distinguishing solid and dotted portions of Stokes curves

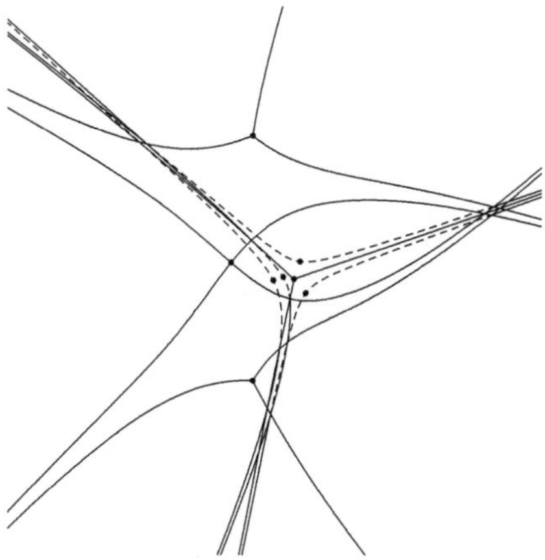

Fig. 3.9 The Stokes geometry obtained from Fig. 3.5 by distinguishing solid and dotted portions of Stokes curves

- A Stokes curve emanating from a virtual turning point is a dotted line near that virtual turning point.
- A Stokes curve changes from a dotted line to a solid line when it passes through an ordered crossing point.

Further, if all portions of a Stokes curve are inert (i.e., represented by dotted lines), we delete such a redundant Stokes curve and its virtual turning point. Figure 3.8 shows the Stokes geometry obtained from Fig. 3.4 by distinguishing solid and dotted portions of each Stokes curve. We can observe that ordered crossing points still exist in Fig. 3.8.

Similarly, Fig. 3.9 shows the Stokes geometry obtained from Fig. 3.5 by the same procedure. One interesting and important point is that, if we apply the same procedure to Figs. 3.6 and 3.7, the resulting Stokes geometries are the same with Fig. 3.9. In other words, the Stokes geometry is saturated for $M = 1$ and all the virtual turning points newly obtained for $M \geq 2$ are redundant. We can also confirm that there are no ordered crossing points in Fig. 3.9.

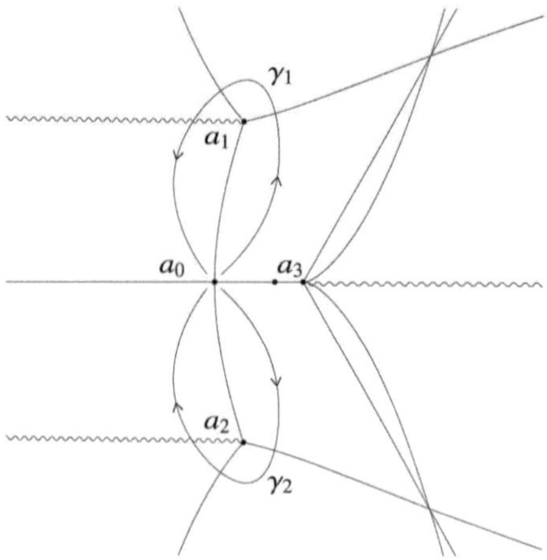

Fig. 3.10 The Stokes geometry of $\widetilde{\mathcal{N}(c)}$ for $c = 1$ and the curves γ_1 and γ_2

In the Appendix we use the exact steepest descent method to confirm whether Stokes phenomena do really occur or not on each portion of Stokes curves.

We now study how the period discussed in the previous section is related with the Stokes geometry of $\widetilde{\mathcal{N}(c)}$. For that purpose let us consider the case where $c = 1$. In this case the period ϖ is real (its exact value is $-1/(3\sqrt{3})$) and the Stokes geometry of $\widetilde{\mathcal{N}(c)}$ becomes the one shown in Fig. 3.10. Note that (3.1.2) or (3.1.3) when $c = 1$ implies

$$\varpi = \int_{\gamma_1} \zeta(z)dz = \int_{\gamma_2} \zeta(z)dz = -\frac{1}{3\sqrt{3}}, \tag{3.2.1}$$

where γ_1 and γ_2 are the curves shown in Fig. 3.10. In this case the type of the non-hereditary double turning point a_0 is the same as those of the simple turning points a_1 and a_2 and, as a consequence of (3.2.1), the turning points a_1 and a_2 are both connected with a_0 by Stokes curves. Otherwise stated, the reality of the period ϖ causes the Stokes geometry of $\widetilde{\mathcal{N}(c)}$ shown in Fig. 3.10 to be a degenerate one. We also note that in this case the type of another simple turning point a_3 is different from that of a_0. Thus the turning points a_3 and a_0 are not connected by a Stokes curve of the same type and there is no degeneration with the portion of the Stokes geometry of $\widetilde{\mathcal{N}(c)}$ containing a_3 and a_0, although in Fig. 3.10 these two turning points seem to be connected by a Stokes curve. (Precisely speaking, two Stokes curves of different types are overlapping on the interval $[a_0, a_3]$.)

In conclusion, for the perturbed system $\widetilde{\mathcal{N}(c)}$ of the tangential system $\mathcal{N}(c)$, we have observed the following intriguing properties:

(i) Due to the existence of the non-hereditary double turning point the singularities of solutions of the Borel transform of $\widetilde{\mathcal{N}(c)}$ propagate along the bicharacteristic chain.
(ii) As its consequence, the singularities have a periodic structure.

(iii) This periodic structure produces many additional virtual turning points of $\widetilde{\mathcal{N}(c)}$ besides virtual turning points of the original system $\mathcal{N}(c)$.
(iv) Most of these additional virtual turning points are redundant.
(v) The period is also related with degeneration of the Stokes geometry of $\widetilde{\mathcal{N}(c)}$.

3.3 Stokes Geometry Related to the (1, 4) Hypergeometric System: Typical Examples of Degeneration When Some Periods Are Real

In this section, using the same arguments as in Sects. 3.1 and 3.2 for $\mathcal{N}(c)$, we consider the Stokes geometry of $\mathcal{N}_{(1,4)HG}[x_1 = c]$ which we have discussed in Sect. 1.5.5. In particular, we investigate the relationship between the Stokes geometry and the periods for $\mathcal{N}_{(1,4)HG}[x_1 = c]$ in a more detailed manner.

The explicit form of $\mathcal{N}_{(1,4)HG}[x_1 = c]$ is given by (1.5.7), which is equivalent to the following third-order differential equation:

$$P^{(\alpha,c)}\psi = 0$$

with

$$P^{(\alpha,c)} = 9\left(\eta^{-1}\frac{d}{dx}\right)^3 + (-4x^2+6c)\left(\eta^{-1}\frac{d}{dx}\right)^2 + (4\alpha x + c^2)\eta^{-1}\frac{d}{dx} - \alpha^2 + Q,$$

where $Q = \sum_{j\geq 1}\eta^{-j}q_j(x,\eta^{-1}d/dx)$ consists of **WKB-theoretically lower-order terms**. In what follows the **WKB-theoretically principal part** of $P^{(\alpha,c)}$ is denoted by $P_0^{(\alpha,c)}$. As was already computed in Sect. 1.5.5, $\mathcal{N}_{(1,4)HG}[x_1 = c]$ has simple turning points at zeros of

$$32\alpha x^3 + 4c^2x^2 - 108\alpha c x - 243\alpha^2 - 12c^3$$

and a non-hereditary double turning point at $x = -(3\alpha)/(2c)$. Similarly to the case of $\mathcal{N}(c)$ (cf. (1.4.3)), the principal symbol $p_0^{(\alpha,c)}(x,\xi,\eta)$ of the Borel transform of $P_0^{(\alpha,c)}$ has the factorization

$$p_0^{(\alpha,c)}(x,\xi,\eta) = q_+(x,\xi,\eta)q_-(x,\xi,\eta)$$

where

$$q_\pm(x,\xi,\eta) = (3\eta^{-3/2}\xi^{3/2} + c\eta^{-1/2}\xi^{1/2}) \pm (2x\eta^{-1}\xi - \alpha).$$

Hence we can argue $\mathscr{N}_{(1,4)HG}[x_1 = c]$ in a parallel way to $\mathscr{N}(c)$. To be more specific, the bicharacteristic strip of $q_+(x, \xi, \eta)$, that is, a solution of the hamiltonian system

$$\begin{cases} \dfrac{dx}{dt} = \dfrac{\partial q_+}{\partial \xi} = \dfrac{9}{2}\eta^{-3/2}\xi^{1/2} + \dfrac{c}{2}\eta^{-1/2}\xi^{-1/2} + 2x\eta^{-1}, \\ \dfrac{dy}{dt} = \dfrac{\partial q_+}{\partial \eta} = -\dfrac{9}{2}\eta^{-5/2}\xi^{3/2} - \dfrac{c}{2}\eta^{-3/2}\xi^{1/2} - 2x\eta^{-2}\xi, \\ \dfrac{d\xi}{dt} = -\dfrac{\partial q_+}{\partial x} = -2\eta^{-1}\xi, \\ \dfrac{d\eta}{dt} = -\dfrac{\partial q_+}{\partial y} = 0 \end{cases}$$

with the initial condition

$$(x(0), y(0), \xi(0), \eta(0)) = \left(-\frac{3\alpha}{2c}, 0, -\frac{c}{3}, 1\right)$$

is given by

$$\mathscr{b}_+ : \quad (x_+(\sigma), y_+(\sigma), \xi_+(\sigma), \eta_+(\sigma)) = (\tilde{x}(\sigma), \tilde{y}(\sigma) - \tilde{y}(0), \tilde{\xi}(\sigma), 1),$$

where

$$\begin{aligned} \tilde{x}(\sigma) &= -\frac{3\alpha}{2c}e^{2\sigma} + \frac{3}{2}\left(-\frac{c}{3}\right)^{1/2} e^{\sigma} - \frac{3}{2}\left(-\frac{c}{3}\right)^{1/2} e^{-\sigma}, \\ \tilde{y}(\sigma) &= \frac{3}{2}\left(-\frac{c}{3}\right)^{3/2} e^{-\sigma} + \frac{1}{2}\left(-\frac{c}{3}\right)^{3/2} e^{-3\sigma} - \alpha\sigma, \\ \tilde{\xi}(\sigma) &= -\frac{c}{3}e^{-2\sigma}. \end{aligned}$$

On the other hand, the bicharacteristic strip of $q_-(x, \xi, \eta)$ with the initial condition

$$(x(0), y(0), \xi(0), \eta(0)) = \left(-\frac{3\alpha}{2c}, 0, -\frac{c}{3}, 1\right)$$

is given by

$$\begin{aligned} \mathscr{b}_- : \quad & (x_-(\sigma), y_-(\sigma), \xi_-(\sigma), \eta_-(\sigma)) \\ &= (x_+(-\sigma - \pi i), y_+(-\sigma - \pi i) - \varpi', \xi_+(-\sigma - \pi i), 1), \end{aligned}$$

where

$$\varpi' = \pi i\alpha - 4\left(-\frac{c}{3}\right)^{3/2}.$$

These explicit formulas for the bicharacteristic strips imply, after a simple change of the time variable σ, that $\mathscr{b}_-$ can be considered as the bicharacteristic strip of $q_+(x, \xi, \eta)$ with the initial condition

$$\left(-\frac{3\alpha}{2c}, -\varpi', -\frac{c}{3}, 1\right)$$

and $\mathscr{b}_+$ can be considered as the bicharacteristic strip of $q_-(x, \xi, \eta)$ with the initial condition

$$\left(-\frac{3\alpha}{2c}, \varpi', -\frac{c}{3}, 1\right).$$

Additionally, since we have

$$x_+\Big|_{\sigma=0} = x_-\Big|_{\sigma=0}, \quad y_+\Big|_{\sigma=0} = y_-\Big|_{\sigma=0},$$

$$\frac{dy_+}{d\sigma} \Big/ \frac{dx_+}{d\sigma}\Big|_{\sigma=0} = \frac{dy_-}{d\sigma} \Big/ \frac{dx_-}{d\sigma}\Big|_{\sigma=0},$$

the bicharacteristic curves $\pi(\mathscr{b}_+)$ and $\pi(\mathscr{b}_-)$ are mutually tangent at $(x, y) = (x(0), y(0)) = (-(3\alpha)/(2c), 0)$. In addition to the above, we consider the bicharacteristic strip $\mathscr{b}_m$ of $q_+(x, \xi, \eta)$ with the initial condition

$$\left(-\frac{3\alpha}{2c}, m\varpi', -\frac{c}{3}, 1\right)$$

for $m \in \mathbb{Z}$, which is explicitly given by

$$\begin{aligned} \mathscr{b}_m : \quad (x_m(\sigma), y_m(\sigma), \xi_m(\sigma), \eta_m(\sigma)) &= (x_+(\sigma), y_+(\sigma) + m\varpi', \xi_+(\sigma), \eta_+(\sigma)) \\ &= (\tilde{x}(\sigma), \tilde{y}(\sigma) - \tilde{y}(0) + m\varpi', \tilde{\xi}(\sigma), 1). \end{aligned}$$

As the argument in Sect. 3.1 shows, singularities of solutions of the Borel transform of $\mathcal{N}_{(1,4)HG}[x_1 = c]$ propagate along the bicharacteristic chain, that is, the collection of $\pi(\mathscr{b}_m)$ with the parametrization

$$\pi(\mathscr{b}_m) : \quad (x_m(\sigma), y_m(\sigma)) = (\tilde{x}(\sigma), \tilde{y}(\sigma) - \tilde{y}(0) + m\varpi') \quad (m \in \mathbb{Z}).$$

Furthermore, $y_m(\sigma + 2\pi i) = y_m(\sigma) - 2\pi i\alpha$ holds. Taking all these facts into account, we conclude that the singularities of the Borel transform of $\mathcal{N}_{(1,4)HG}[x_1 = c]$ have a periodic structure described by $\varpi = 2\pi i\alpha$ and $\varpi' = \pi i\alpha - 4(-c/3)^{3/2}$.

The constants ϖ and ϖ' are expressed in the form of contour integrals. For example, ϖ is given by an integral

$$\varpi = \int_\gamma \xi(x)dx,$$

where $\xi(x)$ is a root of $p_0^{(\alpha,c)}(x, \xi, 1) = 0$ that does not have a branch point at $x = \infty$, and γ is a small closed curve around $x = \infty$. In fact, letting $w = 1/x$, we find $\xi(x)$ can be expanded near $w = 0$ as

$$\xi(x) = \frac{4}{9w^2} - \frac{2c}{3} - \alpha w + \cdots$$

and hence

$$\begin{aligned}\int_\gamma \xi(x)dx &= \int_{\text{around } w=0} \left(\frac{4}{9w^2} - \frac{2c}{3} - \alpha w + \cdots\right)\left(-\frac{1}{w^2}\right) dw \\ &= 2\pi i\alpha \\ &= \varpi.\end{aligned}$$

The other period ϖ' is also given by an integral

$$\varpi' = \int_{\gamma'} \xi(x)dx,$$

where γ' denotes a closed curve on the Riemann surface of the root $\xi(x)$ of $p_0^{(\alpha,c)}(x, \xi, 1) = 0$ which is defined in a similar manner to the curve γ of the expression (3.1.2) discussed in Sect. 3.1. That is, γ' is defined by $x = x_+(\sigma(t))$ $(0 \le t \le 1)$, where $\sigma(t)$ $(0 \le t \le 1)$ is a curve in the σ-plane satisfying

$$\sigma(0) = \pi i, \quad \sigma(1) = 2\pi i$$

and

$$\sigma(t) \in \mathbb{C} \setminus \left\{\sigma^{(j)} + 2\pi i m\ ;\ 1 \le j \le 3, m \in \mathbb{Z}\right\} \quad (0 < t < 1).$$

Here $\sigma^{(j)} + 2\pi i m$ $(1 \le j \le 3, m \in \mathbb{Z})$ is a solution of $(dx_+/d\sigma)(\sigma) = 0$ and hence $x_+(\sigma^{(j)} + 2\pi i m)$ is a simple turning point of $\mathcal{N}_{(1,4)HG}[x_1 = c]$. On the other hand, $\sigma = \pi i m$ $(m \in \mathbb{Z})$ corresponds to the non-hereditary double turning point of $\mathcal{N}_{(1,4)HG}[x_1 = c]$ since

$$(x_+(\sigma), \xi_+(\sigma)) = \left(-\frac{3\alpha}{2c}, -\frac{c}{3}\right),$$

holds at $\sigma = \pi i m$. Consequently the starting point and endpoint of γ' are both the non-hereditary double turning point $-(3\alpha)/(2c)$. Furthermore, on the curve γ' an appropriate branch $\xi = \xi(x)$ of the root of $p_0^{(\alpha,c)}(x, \xi, 1) = 0$ is chosen so that $\xi_+(\sigma) = \xi(x_+(\sigma))$ holds. Using this curve γ', we can confirm through calculations similar to those in Sect. 3.1 that

$$\begin{aligned}\int_{\gamma'} \xi(x)dx &= -\tilde{y}(2\pi i) + \tilde{y}(\pi i) \\ &= \pi i \alpha - 4\left(-\frac{c}{3}\right)^{3/2} \\ &= \varpi'.\end{aligned}$$

Based on these observations, we now consider the relationship between the periods and the Stokes geometry. We first recall the case of $(c_1, \alpha) = (c_{10}, \alpha_0) = (-3/4, 2i)$ discussed in Sect. 1.5.5. In this case the periods are $\varpi = -4\pi, \varpi' = -2\pi - 1/2$, both of which are real. The Stokes geometry of $\mathcal{N}_{(1,4)HG}[x_1 = c_1]$ for $(c_1, \alpha) = (c_{10}, \alpha_0)$ is given in Fig. 3.11 (which is the same as Fig. 1.13). It is readily observed that all simple turning points a_1, a_2 and a_3 are connected with the non-hereditary double turning point by Stokes curves. This degenerate configuration of the Stokes geometry is an effect of the reality of the periods ϖ and ϖ'.

Next, adding some small perturbations to $(c_1, \alpha) = (c_{10}, \alpha_0)$, we consider the following three cases:

$$(c_1, \alpha) = (c_{10}, \alpha_0) + (\kappa_1, \lambda_1), \quad (c_{10}, \alpha_0) + (\kappa_2, \lambda_2), \quad (c_{10}, \alpha_0) + (\kappa_1, \lambda_1) + (\kappa_2, \lambda_2),$$

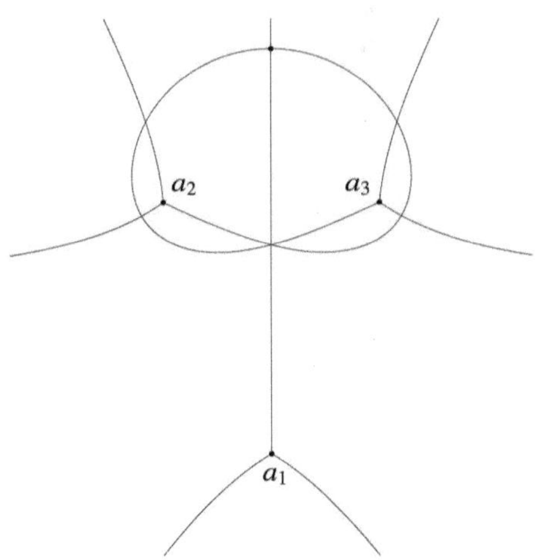

Fig. 3.11 The Stokes geometry of $\mathcal{N}_{(1,4)HG}[x_1 = c_1]$ for $(c_1, \alpha) = (c_{10}, \alpha_0)$

where the perturbations (κ_1, λ_1) and (κ_2, λ_2) are assumed to satisfy

$$\varpi|_{(c_1,\alpha)=(c_{10},\alpha_0)+(\kappa_1,\lambda_1)} = \varpi|_{(c_1,\alpha)=(c_{10},\alpha_0)} \in \mathbb{R}, \quad \varpi'|_{(c_1,\alpha)=(c_{10},\alpha_0)+(\kappa_1,\lambda_1)} \notin \mathbb{R},$$

$$\varpi|_{(c_1,\alpha)=(c_{10},\alpha_0)+(\kappa_2,\lambda_2)} \notin \mathbb{R}, \quad \varpi'|_{(c_1,\alpha)=(c_{10},\alpha_0)+(\kappa_2,\lambda_2)} = \varpi'|_{(c_1,\alpha)=(c_{10},\alpha_0)} \in \mathbb{R},$$

$$\varpi|_{(c_1,\alpha)=(c_{10},\alpha_0)+(\kappa_1,\lambda_1)+(\kappa_2,\lambda_2)} \notin \mathbb{R}, \quad \varpi'|_{(c_1,\alpha)=(c_{10},\alpha_0)+(\kappa_1,\lambda_1)+(\kappa_2,\lambda_2)} \notin \mathbb{R}.$$

For example, if we take $(\kappa_1, \lambda_1) = ((3/100)i, 0)$ and $(\kappa_2, \lambda_2) = ((3/100)i, \varepsilon)$, where ε is a solution of

$$\varpi'|_{(c_1,\alpha)=(c_{10},\alpha_0)+(\frac{3}{100}i,\varepsilon)} = \varpi'|_{(c_1,\alpha)=(c_{10},\alpha_0)},$$

that is,

$$\varepsilon = \frac{4}{\pi i}\left\{\left(-\frac{c_{10}+(3/100)i}{3}\right)^{3/2} - \left(-\frac{c_{10}}{3}\right)^{3/2}\right\},$$

then the above conditions are satisfied. The Stokes geometries for $(c_1, \alpha) = (c_{10}, \alpha_0) + (\kappa_1, \lambda_1)$, $(c_{10}, \alpha_0) + (\kappa_2, \lambda_2)$, and $(c_{10}, \alpha_0) + (\kappa_1, \lambda_1) + (\kappa_2, \lambda_2)$ are shown in Figs. 3.12, 3.13, and 3.14, respectively. In these figures, we observe the following:

- For $(c_1, \alpha) = (c_{10}, \alpha_0) + (\kappa_1, \lambda_1)$, two simple turning points a_1 and a_3 are connected by a Stokes curve.
- For $(c_1, \alpha) = (c_{10}, \alpha_0) + (\kappa_2, \lambda_2)$, a simple turning point a_1 and the non-hereditary double turning point are connected by a Stokes curve.
- For $(c_1, \alpha) = (c_{10}, \alpha_0) + (\kappa_1, \lambda_1) + (\kappa_2, \lambda_2)$, none of the two turning points is connected by a Stokes curve.

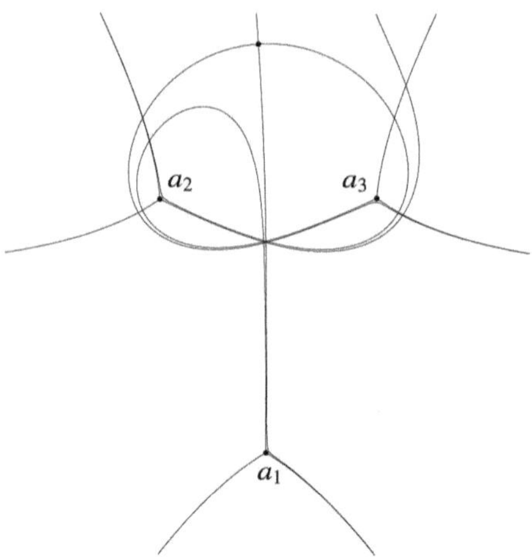

Fig. 3.12 The Stokes geometry of $\mathcal{N}_{(1,4)HG}[x_1 = c_1]$ for $(c_1, \alpha) = (c_{10}, \alpha_0) + (\kappa_1, \lambda_1)$

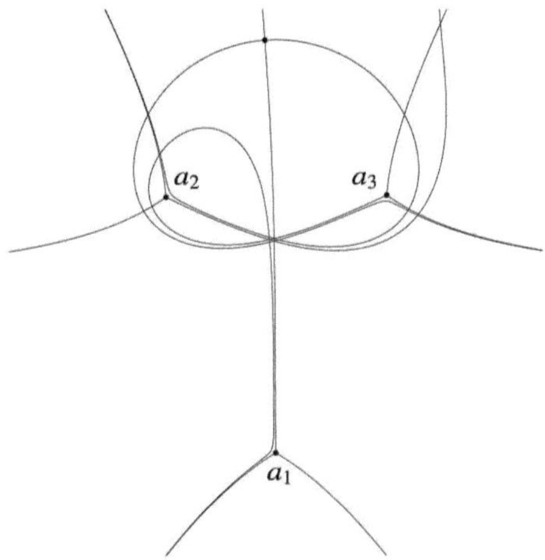

Fig. 3.13 The Stokes geometry of $\mathcal{N}_{(1,4)HG}[x_1 = c_1]$ for $(c_1, \alpha) = (c_{10}, \alpha_0) + (\kappa_2, \lambda_2)$

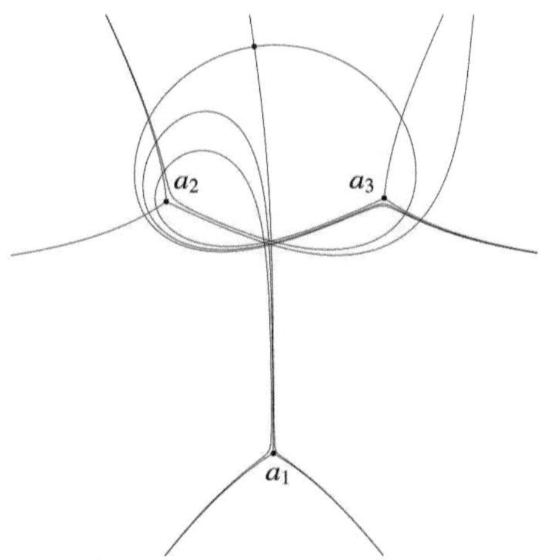

Fig. 3.14 The Stokes geometry of $\mathcal{N}_{(1,4)HG}[x_1 = c_1]$ for $(c_1, \alpha) = (c_{10}, \alpha_0) + (\kappa_1, \lambda_1) + (\kappa_2, \lambda_2)$

In particular, Fig. 3.15 is an enlarged version of some relevant portion of Fig. 3.12 (non-relevant Stokes curves are omitted). Note that Fig. 3.15 contains the tadpole-type configuration, that is, there exists a crossing point p of three Stokes curves and two of these three crossing Stokes curves are the same Stokes curve. The simple turning points a_1 and a_3 are connected by this Stokes curve in Fig. 3.15.

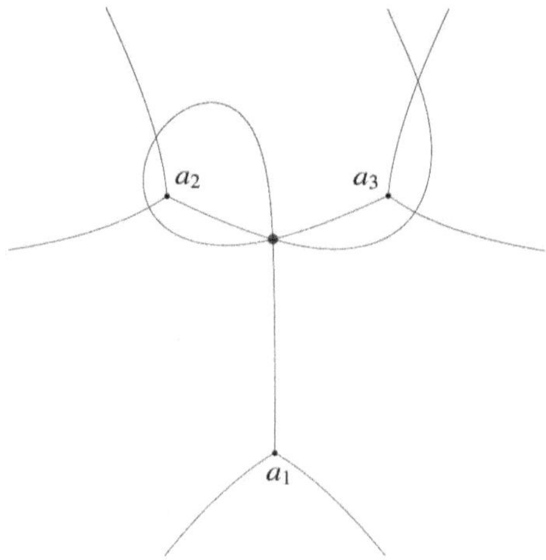

Fig. 3.15 Some relevant portion of Fig. 3.12 enlarged. The red point is the crossing point p in question

3.4 A Connection Formula for a Non-hereditary Turning Point Related to a Leaf-Type and a Tadpole-Type Configurations

As we have seen so far, particularly in Sect. 1.5 and Chap. 2, two types (i.e., trilemma-type and triordered-type) of crossing phenomena of three Stokes curves are commonly observed with the configuration of Stokes curves for higher-order equations. As their particular cases, we also encounter leaf-type configurations and tadpole-type configurations in conjunction with non-hereditary turning points. In this final section, using the fundamental connection formulas for Borel resummed WKB solutions, we show that leaf-type crossing points and tadpole-type crossing points are both of triordered type and, in particular, that one of the six portions of Stokes curves around a tadpole-type crossing point is inert.

We first discuss tadpole-type configurations. To make the situation concrete, let us consider a tadpole-type configuration we encountered in Sect. 1.5. We may assume that a Stokes curve emanating from a simple turning point $x = a$ on the right is labeled as $2 < 3$ without loss of generality. We now assume that a simple turning point $x = \tilde{a}$ on the left is labeled as $(1, 2)$ or $1 = 2$ and, temporarily forgetting a Stokes curve emanating from $\tilde{a}$, we let P denote a self-intersection point of the Stokes curve emanating from a. Then, since a real-valued function $\mathrm{Re} \int^x (\zeta_j - \zeta_k) dx$ is monotonically increasing or decreasing on each Stokes curve defined by $\mathrm{Im} \int^x (\zeta_j - \zeta_k) dx = 0$, we find

$$\int_{a,\ \text{along}\ \Gamma_1 \cup \Gamma_2}^{P} (\zeta_3 - \zeta_2)\, dx > \int_{a,\ \text{along}\ \Gamma_1}^{P} (\zeta_3 - \zeta_2)\, dx > 0, \tag{3.4.1}$$

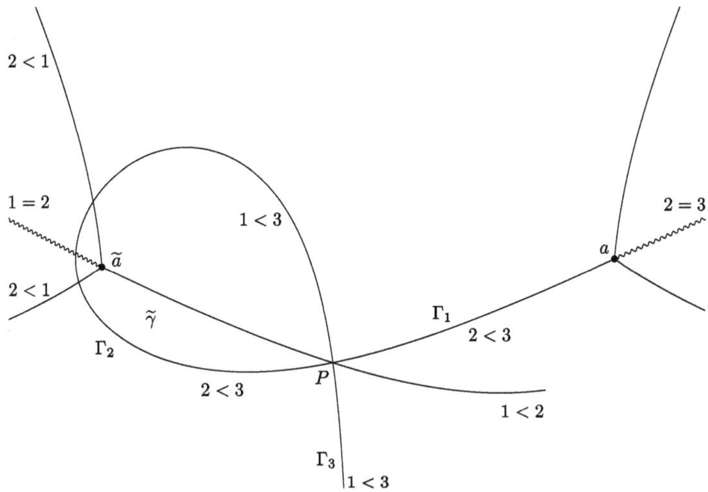

Fig. 3.16 Stokes curves around a tadpole-type crossing point P

where $\zeta_j = \zeta_j(x)$ $(j = 1, 2, 3)$ denotes an appropriately labeled characteristic root of the differential equation in question on this cut plane, Γ_1 designates the portion of the Stokes curve between a and P, Γ_2 designates the portion of the Stokes curve between P and P encircling the turning point $\tilde{a}$, and Γ_3 designates the portion of the same Stokes curve after it passes through P twice (cf. Fig. 3.16). Note that ζ_2 changes its label to ζ_1 after the Stokes curve Γ_2 crosses the cut drawn from $\tilde{a}$. It then follows from (3.4.1) that

$$\int_{P,\ \text{along}\ \Gamma_2}^{P} (\zeta_3 - \zeta_2)\, dx > 0,$$

and hence, as $\displaystyle\int_{P,\ \text{along}\ \Gamma_2}^{P} \zeta_3 dx = 0$, we have

$$\int_{P,\ \text{along}\ \Gamma_2}^{P} (-\zeta_2)\, dx > 0,$$

that is,

$$\int_{\tilde{a}}^{P} (\zeta_2 - \zeta_1)\, dx > 0. \tag{3.4.2}$$

Inequality (3.4.2) confirms that the crossing point P is inevitably located on a Stokes curve emanating from $\tilde{a}$ and that the label of this Stokes curve emanating from $\tilde{a}$ is $2 > 1$. In particular, the tadpole-type crossing point P is of triordered type.

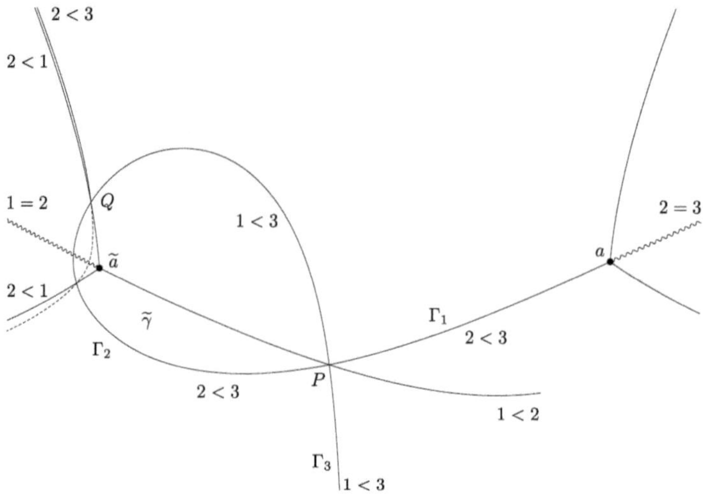

Fig. 3.17 Stokes curves of Fig. 3.16 with relevant new Stokes curves being added

We next show that the portion Γ_3 of the Stokes curve is inert. To this end, we add a relevant (new) Stokes curve to Fig. 3.16 so that the computation of **connection coefficients** based on the configuration of Stokes curves becomes complete. The result is given in Fig. 3.17. In Fig. 3.17 a (new) Stokes curve passing through a crossing point Q of Γ_2 with a Stokes curve emanating from $\tilde{a}$ is added since Q is an ordered crossing point. As a portion of a Stokes curve containing a virtual turning point is inert, the portion of this newly added Stokes curve containing another crossing point of it with Γ_2 is expected to be inert and designated by a dotted line in Fig. 3.17. Note that, when we encounter a tadpole-type configuration, a Stokes curve encircling a simple turning point always crosses two Stokes curves emanating from that turning point in addition to the one constituting a tadpole-type configuration. Furthermore, one of these two crossing points is inevitably an ordered crossing point and a (new) Stokes curve passing through this ordered crossing point should be added. In what follows, after the relevant (new) Stokes curve is added, the configuration is assumed to be the one given in Fig. 3.17. Now, taking a base point x_0 as in Fig. 3.18, we consider a WKB solution ψ_j $(j = 1, 2, 3)$ normalized as

$$\psi_j = \exp \int_{x_0}^{x} S^{(j)}(x, \eta)\, dx = \exp \int_{x_0}^{x} (\eta \zeta_j(x) + \cdots)\, dx. \tag{3.4.3}$$

Here, and in what follows, $S^{(j)}(x, \eta)$ denotes a solution of the corresponding higher-order version of the Riccati equation with the choice of $S_{-1}(x) = \zeta_j(x)$. We also let c_k $(k = 1, 2, 3, 4)$ denote the connection coefficient attached to each portion of Stokes curves shown in Fig. 3.18. That is, if c_k is attached to a portion of a Stokes curve of type $j_1 > j_2$, a Stokes phenomenon $\psi_{j_1} \mapsto \psi_{j_1} + c_k \psi_{j_2}$ occurs when the Borel sum of ψ_{j_1} is analytically continued across the portion of the Stokes curve in

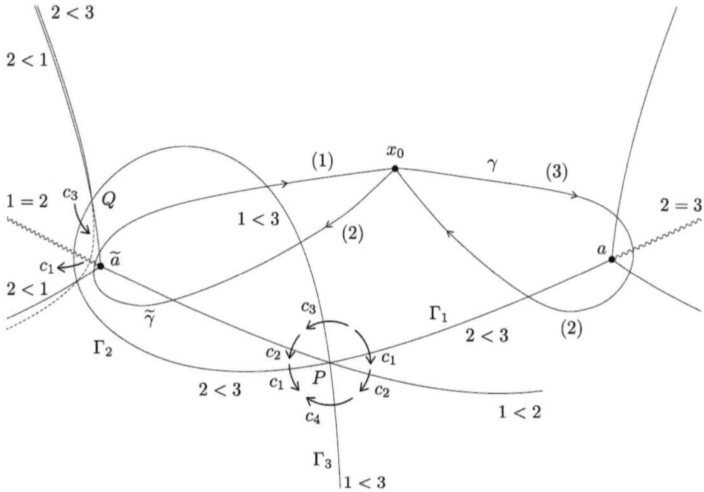

Fig. 3.18 Stokes curves of Fig. 3.17 with a base point x_0 being added

question in the direction indicated by an arrow in Fig. 3.18. Note that, since P is a triordered-type crossing point as verified above, the connection coefficients c_3 and c_4 attached to both sides of a non-adjacent Stokes curve of type $1 < 3$ are different in general. In fact, Recipe 1.5.1 of [VTP] (cf. [VTP, Formula (1.5.2)]) assures the following relation:

$$c_4 = c_3 - c_1 c_2. \tag{3.4.4}$$

On the other hand, c_1 and c_2 attached to adjacent Stokes curves respectively of type $2 < 3$ and $1 < 2$ are unchanged at P. Similarly, the connection coefficient c_3 is unchanged at another ordered crossing point Q. Taking account of these properties for c_k, we explicitly compute c_1, c_2 and c_3 to confirm the vanishing of c_4 in what follows.

Since c_1 is attached to a Stokes curve of type $2 < 3$ emanating from a simple turning point a, it is readily computed by applying the argument of [VTP, Section 1.2]. As a matter of fact, if a is a simple turning point of type (j, k) and if we let $\gamma_{k \to j}$ (resp., $\gamma_{j \to k}$) denote a path on the Riemann surface of the characteristic root $\zeta(x)$ that emanates from a base point x_0 on the k-th (resp., j-th) sheet, encircles a in a clockwise manner and returns to x_0 on the j-th (resp., k-th) sheet as is shown in Fig. 3.19, then the following connection formula holds across a Stokes curve of type $j < k$:

$$\psi_k \mapsto \psi_k + c\psi_j \tag{3.4.5}$$

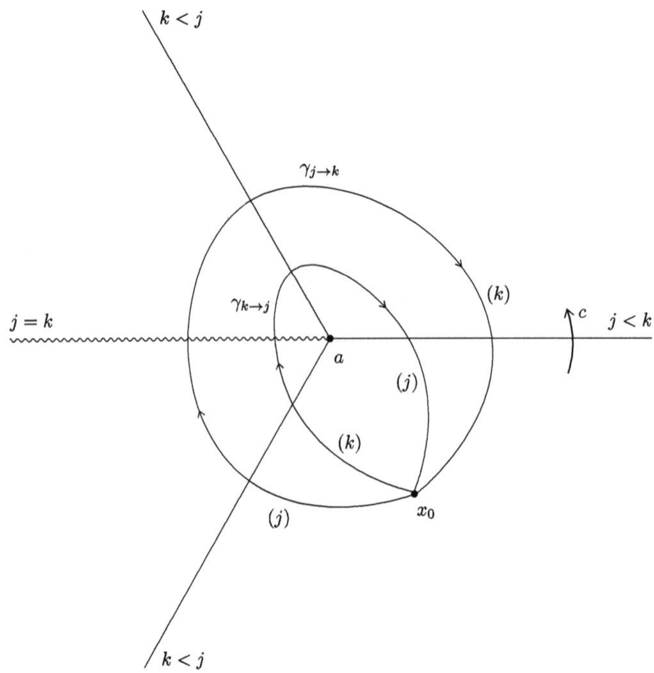

Fig. 3.19 Paths $\gamma_{j\to k}$ and $\gamma_{k\to j}$

with

$$c = \exp\int_{\gamma_{k\to j}} S(x,\eta)\,dx = -\exp\left(-\int_{\gamma_{j\to k}} S(x,\eta)\,dx\right), \tag{3.4.6}$$

where ψ_l $(l = j, k)$ is a WKB solution normalized as follows:

$$\psi_l(x,\eta) = \exp\int_{x_0}^{x} S^{(l)}(x,\eta)\,dx = \exp\int_{x_0}^{x} (\eta\zeta_l(x) + \cdots)\,dx. \tag{3.4.7}$$

(Cf. [V], [KT] and [Sa1]. In the case where one crosses a Stokes curve in an opposite direction, that is, in a clockwise manner viewed from a simple turning point a, c should be replaced by $-c$.) Using the connection formula (3.4.5), we find

$$c_1 = \exp\int_{\gamma} S(x,\eta)\,dx, \tag{3.4.8}$$

where γ is a path on the Riemann surface of $\zeta(x)$ indicated in Fig. 3.18. Similarly, we have

$$c_2 = -\exp \int_{\tilde{\gamma}} S(x, \eta)\, dx \tag{3.4.9}$$

with $\tilde{\gamma}$ being a path also indicated in Fig. 3.18. Finally, let us compute c_3. The connection coefficients c_3 and $-c_1$ are basically the same as both of them are attached to the same Stokes curve Γ_2. However, as Γ_2 encircles the simple turning point $\tilde{a}$, the normalization of WKB solutions becomes different when we consider the corresponding Stokes phenomenon near the crossing point P and consequently c_3 and $-c_1$ have different expressions. To be more specific, if one crosses the Stokes curve Γ_2 near P in a direction indicated by an arrow of c_3 in Fig. 3.18, we have the following connection formula:

$$\tilde{\psi}_3 \mapsto \tilde{\psi}_3 + (-c_1)\tilde{\psi}_2, \tag{3.4.10}$$

where a WKB solution $\tilde{\psi}_3$ should be

$$\tilde{\psi}_3 = \exp\left(\int_{x_0}^{P} S^{(3)} dx + \int_{\Gamma_2} S^{(3)} dx + \int_{P}^{x} S^{(3)} dx\right) = \exp \int_{x_0}^{x} S^{(3)} dx = \psi_3 \tag{3.4.11}$$

and $\tilde{\psi}_2$ should be

$$\begin{aligned} \tilde{\psi}_2 &= \exp\left(\int_{x_0}^{P} S^{(2)} dx + \int_{\Gamma_2} S^{(2)\to(1)} dx + \int_{P}^{x} S^{(1)} dx\right) \\ &= \exp\left(\int_{\tilde{\gamma}} S dx + \int_{x_0}^{x} S^{(1)} dx\right) = \exp\left(\int_{\tilde{\gamma}} S dx\right)\psi_1. \end{aligned} \tag{3.4.12}$$

Hence the connection formula (3.4.10) can be written also as

$$\psi_3 \mapsto \psi_3 - c_1 \exp\left(\int_{\tilde{\gamma}} S dx\right)\psi_1, \tag{3.4.13}$$

which exactly means that

$$c_3 = -c_1 \exp\left(\int_{\tilde{\gamma}} S dx\right). \tag{3.4.14}$$

Combining (3.4.9) and (3.4.14), we obtain

$$c_3 = c_1 c_2. \tag{3.4.15}$$

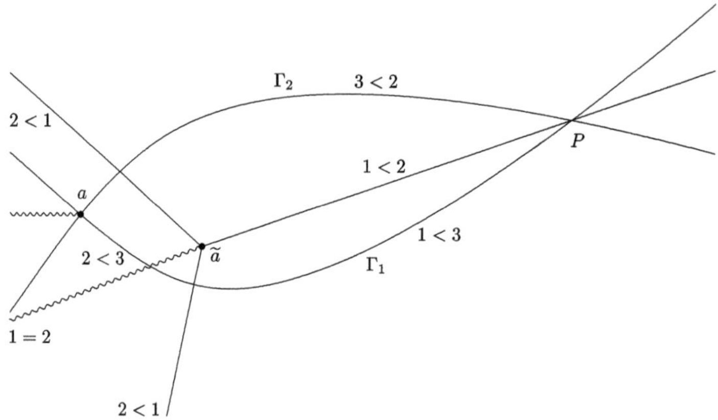

Fig. 3.20 Stokes curves around a leaf-type crossing point P

This together with (3.4.4) implies

$$c_4 = 0. \tag{3.4.16}$$

We have thus verified that the Stokes curve Γ_3 is inert, otherwise stated, no Stokes phenomenon occurs on the Stokes curve Γ_3.

Next, we consider leaf-type configurations. The case of leaf-type configurations is discussed in a way parallel to the case of tadpole-type configurations. To make the situation concrete, we again deal with a leaf-type configuration we encountered in Sect. 1.5. We assume that Stokes curves Γ_1 and Γ_2 emanating from a non-hereditary double turning point $x = a$ are labeled near a as $2 < 3$ and $3 < 2$, respectively, without loss of generality (cf. Fig. 3.20). Then a simple turning point $x = \tilde{a}$ in the center is labeled as $(1, 2)$ or $1 = 2$ and the label of the Stokes curve Γ_1 is changed to $1 < 3$ after it crosses the cut drawn from $\tilde{a}$.

Now, let us first check that a Stokes curve labeled as $1 < 2$ emanating from $\tilde{a}$ passes through a crossing point P of Γ_1 and Γ_2. In the current situation we have

$$\int_{a,\text{ along }\Gamma_1}^{P} (\zeta_3 - \zeta_2)\, dx > 0, \quad \int_{a,\text{ along }\Gamma_2}^{P} (\zeta_2 - \zeta_3)\, dx > 0. \tag{3.4.17}$$

Summing up these two inequalities, we find

$$\int_{a,\text{ along }\Gamma_1}^{P} (\zeta_3 - \zeta_2)\, dx + \int_{a,\text{ along }\Gamma_2}^{P} (\zeta_2 - \zeta_3)\, dx > 0,$$

that is,

$$\int_{a,\text{ along }\Gamma_2}^{P} \zeta_2\,dx - \int_{a,\text{ along }\Gamma_1}^{P} \zeta_2\,dx > 0.$$

Hence we obtain

$$\int_{\tilde{a}}^{P} (\zeta_2 - \zeta_1)\,dx > 0. \tag{3.4.18}$$

This confirms that the crossing point P is located on a Stokes curve emanating from $\tilde{a}$ with the label $1 < 2$. In particular, the leaf-type crossing point P is of triordered-type.

Finally, we discuss the **connection coefficients** around a leaf-type crossing point P. Similarly to the case of a tadpole-type configuration, consider a WKB solution ψ_j $(j = 1, 2, 3)$ normalized as

$$\psi_j = \exp \int_{x_0}^{x} S^{(j)}(x, \eta)\,dx = \exp \int_{x_0}^{x} (\eta\zeta_j(x) + \cdots)\,dx, \tag{3.4.19}$$

where x_0 is a suitably chosen base point, and let c_k $(1 \le k \le 5)$ denote the connection coefficient attached to each portion of Stokes curves (cf. Fig. 3.21). Around the crossing point P the coefficients c_1 and c_2 attached to adjacent Stokes curves Γ_1 and Γ_2 are unchanged, while c_3 and c_4 attached to a non-adjacent Stokes curve are different. In fact, we have a relation

$$c_4 = c_3 + c_1 c_2 \tag{3.4.20}$$

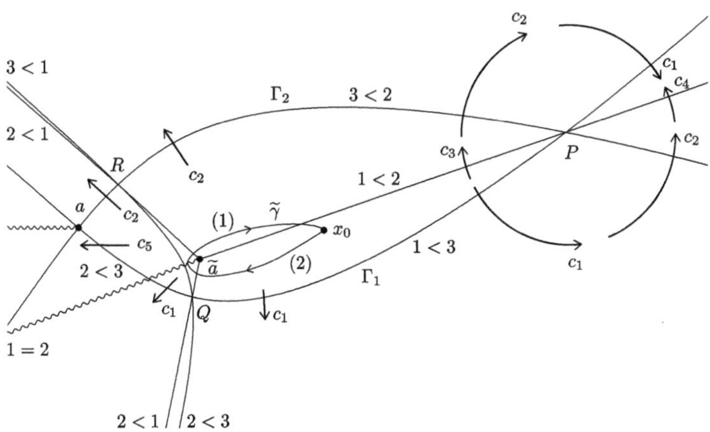

Fig. 3.21 Stokes curves of Fig. 3.20 with a relevant new Stokes curve and a base point being added

in this case. In what follows, through some computations of these coefficients c_j, we try to obtain a more explicit relation between c_3 and c_4. Note that a (new) Stokes curve passing through two ordered crossing points Q and R is added in Fig. 3.21. As in Fig. 3.18, since the Stokes curves Γ_1 and Γ_2 are both adjacent ones at Q and R respectively, the connection coefficients c_1 and c_2 are unchanged there. On the other hand, in Fig. 3.21 there exists a (non-hereditary) double turning point a. As is discussed in [AKT3, Section 1], a second-order differential equation can be transformed to the Weber equation at a double turning point (cf. see also [Sa2], where the Borel summability of the transformation series to the Weber equation is also discussed). Furthermore, the connection formula for the Weber equation with a double turning point is explicitly given in [T2, Section 4]. We omit the detailed discussion here but, combining these results together with the argument of [VTP, Section 1.2] in a manner similar to the case of a simple turning point, we obtain the following connection formula around a double turning point $x = a$ for suitably normalized WKB solutions $\tilde{\psi}_2$ and $\tilde{\psi}_3$:

$$\tilde{\psi}_3 \mapsto \tilde{\psi}_3 - \frac{i\sqrt{2\pi}}{\Gamma(\kappa + 1/2)}\tilde{\psi}_2 \quad \text{on } \Gamma_1, \tag{3.4.21}$$

and

$$\tilde{\psi}_2 \mapsto \tilde{\psi}_2 + \frac{i\sqrt{2\pi}}{\Gamma(-\kappa + 1/2)}e^{i\pi\kappa}\tilde{\psi}_3 \quad \text{on } \Gamma_2, \tag{3.4.22}$$

where

$$\kappa = \operatorname*{Res}_{x=a} \frac{1}{2}(S^{(3)} - S^{(2)}). \tag{3.4.23}$$

Letting M_j denote normalization constants for $\tilde{\psi}_j$ $(j = 2, 3)$, that is,

$$\tilde{\psi}_2 = M_2\psi_2, \quad \tilde{\psi}_3 = M_3\psi_3, \tag{3.4.24}$$

we obtain

$$\psi_3 \mapsto \psi_3 - \frac{i\sqrt{2\pi}}{\Gamma(\kappa + 1/2)}\frac{M_2}{M_3}\psi_2 \quad \text{on } \Gamma_1, \tag{3.4.25}$$

$$\psi_2 \mapsto \psi_2 + \frac{i\sqrt{2\pi}}{\Gamma(-\kappa + 1/2)}e^{i\pi\kappa}\frac{M_3}{M_2}\psi_3 \quad \text{on } \Gamma_2, \tag{3.4.26}$$

which means that

$$c_5 = -\frac{i\sqrt{2\pi}}{\Gamma(\kappa + 1/2)}\frac{M_2}{M_3} \tag{3.4.27}$$

and

$$c_2 = \frac{i\sqrt{2\pi}}{\Gamma(-\kappa + 1/2)} e^{i\pi\kappa} \frac{M_3}{M_2}. \tag{3.4.28}$$

By an argument similar to the derivation of the formula (3.4.14), the connection coefficient c_1 can be determined from (3.4.27) as follows:

$$c_1 = -c_5 \exp\left(\int_{\tilde{\gamma}} S dx\right) = \frac{i\sqrt{2\pi}}{\Gamma(\kappa + 1/2)} \frac{M_2}{M_3} \exp\left(\int_{\tilde{\gamma}} S dx\right), \tag{3.4.29}$$

where $\tilde{\gamma}$ is a path shown in Fig. 3.21. On the other hand, since $\tilde{a}$ is a simple turning point, the connection coefficient c_3 is readily computed as

$$c_3 = \exp \int_{\tilde{\gamma}} S(x, \eta)\, dx \tag{3.4.30}$$

(cf. (3.4.6)). Hence

$$\begin{aligned} c_1 c_2 &= -\frac{2\pi}{\Gamma(\kappa + 1/2)\Gamma(-\kappa + 1/2)} e^{i\pi\kappa} \exp\left(\int_{\tilde{\gamma}} S dx\right) \\ &= -2\cos(\pi\kappa) e^{i\pi\kappa} \exp\left(\int_{\tilde{\gamma}} S dx\right) \\ &= -(e^{2i\pi\kappa} + 1) \exp\left(\int_{\tilde{\gamma}} S dx\right) \\ &= -(e^{2i\pi\kappa} + 1) c_3. \end{aligned} \tag{3.4.31}$$

We thus obtain

$$c_4 = c_3 + c_1 c_2 = -e^{2i\pi\kappa} c_3. \tag{3.4.32}$$

In conclusion, the gap of the connection coefficient at a leaf-type crossing point is described by the residue (3.4.23) of WKB solutions at a non-hereditary turning point $x = a$.

Appendix A
Confirmation of the Activeness of Some Stokes Curves by Using the Exact Steepest Descent Method

A.1 An Overview of the Exact Steepest Descent Method

Following Sects. 1.5.2, 1.5.3 and 3.2, we examine the activeness of some Stokes curves by the **steepest descent method** and the **exact steepest descent method**. For this aim, we briefly review the two methods.

A.1.1 Steepest Descent Method

For the integral with a large parameter η along some path Γ

$$\psi(x, \eta) = \int_\Gamma e^{\eta f(x,\xi)} g(\xi)\, d\xi, \tag{A.1.1}$$

we can employ the steepest descent method to examine the Borel summability of asymptotic expansions (with respect to η) around its saddle points. Here we assume that $f_\xi = \partial f/\partial \xi$ is algebraic. A **saddle point** of ψ is a zero of $f_\xi(x, \xi)$ with respect to ξ, and a **steepest descent path** is a solution curve of $-\operatorname{grad}_{(\mathrm{Re}\xi,\mathrm{Im}\xi)} f(x, \xi)$. In general, if Γ is a steepest descent path passing through a saddle point and the path extends without encountering any other saddle point, the integral has an asymptotic expansion

$$\psi(x, \eta) \sim e^{\eta \xi(x)} \sum_{n=0}^{\infty} \psi_n(x) \eta^{-n-1/2}, \tag{A.1.2}$$

S. Hirose et al., *Virtual Turning Points II*, SpringerBriefs in Mathematical Physics 52, https://doi.org/10.1007/978-981-95-2817-2

where $\xi(x)$ is the value of f at the saddle point, and the integral is its Borel sum. Also, if we consider a differential equation of Laplace type, namely, a differential equation with linear coefficients

$$\left(\frac{d^m}{dx^m} + (c_{m-1}x + d_{m-1})\eta\frac{d^{m-1}}{dx^{m-1}} + \cdots + (c_0x + d_0)\eta^m\right)\psi = 0, \tag{A.1.3}$$

then it has such an integral representation of solutions. The series (A.1.2) turns out to be a WKB solution, and its Borel summability can be examined by checking if the steepest descent path meets a saddle point or not. See [T1] and [VTP, Appendix A.1] for the details.

A.1.2 Exact Steepest Descent Method

For a general equation

$$\left(\sum_{k=0}^{m} a_k(x)\eta^{m-k}\frac{d^k}{dx^k}\right)\psi = 0, \tag{A.1.4}$$

we do not have such a simple expression of solutions as (A.1.1), but we do have one with g replaced by a formal power series $g(\xi, \eta)$. Consider the Eq. (A.1.4) with polynomial coefficients $a_k(x)$. By the Fourier transform $\psi(x, \eta) = \int \hat{\psi}(\xi, \eta)e^{\eta x\xi}\, d\xi$, the equation is transformed into the following one:

$$\left(\sum_{k=0}^{m} a_k(-\eta^{-1}\frac{d}{d\xi})\eta^m\xi^k\right)\hat{\psi} = 0, \tag{A.1.5}$$

Then it has a WKB solution of the form

$$\hat{\psi}(\xi, \eta) = e^{\eta\int x(\xi)\,d\xi}\phi(\xi, \eta) \tag{A.1.6}$$

$$= e^{\eta\int x(\xi)\,d\xi}\sum_{n=0}^{\infty}\phi_n(\xi)\eta^{-(n+1/2)} \tag{A.1.7}$$

and the original equation has a formal solution of the form

$$\psi(x, \eta) = \int e^{\eta x\xi}\hat{\psi}(\xi, \eta)\, d\xi \tag{A.1.8}$$

$$= \int e^{\eta(x\xi+\int x(\xi)\,d\xi)}\phi(\xi, \eta)\, d\xi. \tag{A.1.9}$$

In this case, the integrand $\hat{\psi}(\xi, \eta)$ is a formal solution of the Fourier transformed equation, and we must take into account its Stokes phenomena. If we want to see the Borel summability of a WKB solution $\psi_j(x, \eta)$ labeled with j, first take a saddle point ξ_j (and $\hat{\psi}_k(\xi, \eta)$) corresponding to ψ_j, and pursue the steepest descent path of $\mathrm{Re}\big(x\xi + \int x_k(\xi)\, d\xi\big)$ passing through ξ_j. If it crosses a Stokes curve of (A.1.5) labeled with $k > l$, we pick up the steepest descent path of $\mathrm{Re}\big(x\xi + \int x_l(\xi)\, d\xi\big)$ emanating from the crossing point. We repeat this procedure if the original steepest descent path or any steepest descent path generated as above again crosses a Stokes curve. The totality of steepest descent paths obtained by this process is called an **exact steepest descent path**. It is expected that if the exact steepest descent path extends without hitting any other saddle point, then the corresponding WKB solution is Borel summable. See [AKT2] and [VTP, Appendix A.2] for details.

A.2 Stokes Geometry of $\mathcal{N}_4[x_1 = c_1, x_3 = c_3]$

In this section, following Sect. 1.5.2, we examine the Stokes geometry of $\mathcal{N}_4[x_1 = c_1, x_3 = c_3]$ for $(c_1, c_3) = (\frac{6}{5} + i, -1 - i)$ and $(\frac{6}{5} + i, -i)$ by using the steepest descent method. In particular, we observe that the role of an ordinary Stokes curve is replaced by a new Stokes curve as the value of parameter c_3 changes from $-1-i$ to $-i$.

For $\mathcal{N}_4[x_1 = c_1, x_3 = c_3]$, we have an integral solution

$$\psi(z, \eta) = \int \exp\eta \left(t^5 + c_1 t^3 + x t^2 + c_3 t\right) dt. \tag{A.2.1}$$

The Stokes geometry for $\mathcal{N}_4[x_1 = c_1, x_3 = c_3]$ with $(c_1, c_3) = (\frac{6}{5} + i, -1 - i)$ is as in Fig. A.1. We investigate a neighborhood of the intersection point $p = 4.130956\cdots + i6.724629\cdots$. As shown in the right-hand figure in Fig. A.1, we take 8 values $x = p + 0.3\exp\left(\frac{R\pi i}{180}\right)$ $(R = 31, 45, 61, 75, 205, 226, 240, 256)$. For each value of them, the steepest descent path of $\mathrm{Re}\left(t^5 + c_1 t^3 + x t^2 + c_3 t\right)$ is given in Fig. A.2. Note that between $R = 45$ and $R = 61$, no Stokes phenomenon occurs. Hence the corresponding Stokes curve, which comes from an ordinary simple turning point, is inert after it passes the intersection point, as is expected.

The Stokes geometry for $\mathcal{N}_4[x_1 = c_1, x_3 = c_3]$ with $(c_1, c_3) = (\frac{6}{5} + i, -i)$ is as in Fig. A.3. We investigate a neighborhood of the intersection point $p = 1.788079\cdots + i4.004568\cdots$. As in the right-hand figure in Fig. A.3, we take 8 values $x = p + 0.3\exp\left(\frac{R\pi i}{180}\right)$ $(R = 27, 43, 61, 82, 206, 224, 240, 264)$. For each value of them, the steepest descent path of $\mathrm{Re}\left(t^5 + c_1 t^3 + x t^2 + c_3 t\right)$ is given in Fig. A.4. Again, we can observe that no Stokes phenomenon occurs between $R = 43$ and $R = 61$. Hence the corresponding Stokes curve is inert after it passes the intersection point, as is expected. Also, we see that Stokes phenomena occur between $R = 61$ and $R = 82$ and between $R = 240$ and $R = 264$. Hence, the

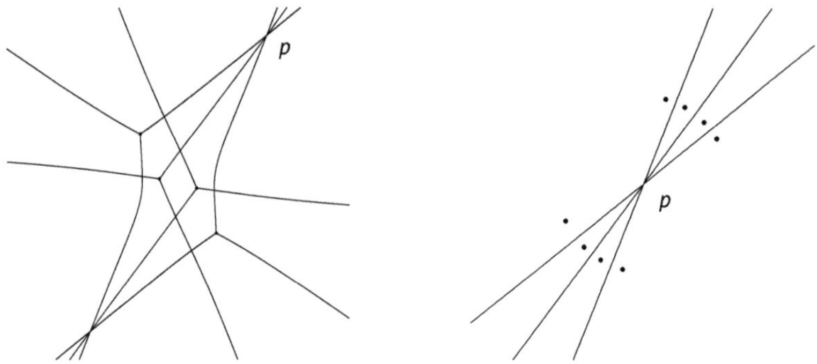

Fig. A.1 Stokes geometry of $\mathcal{N}_4[x_1 = c_1, x_3 = c_3]$ with $(c_1, c_3) = (\frac{6}{5} + i, -1 - i)$. This figure shows the same Stokes geometry as Fig. 1.4, though here inert new Stokes curves are omitted. The right-hand figure is an expansion around the intersection point $p = 4.130956\cdots + i6.724629\cdots$

corresponding new Stokes curve, which is drawn in red in Fig. A.3, is indeed active on both sides of the intersection point p. Thus we have confirmed that, as the value of parameter c_3 changes from $-1 - i$ to $-i$, a new Stokes curve replaces the role of an ordinary Stokes curve.

A.3 Stokes Geometry of $\mathcal{N}(c)$ and $\widetilde{\mathcal{N}(c)}$

In this section, we examine the Stokes geometry of $\mathcal{N}(c)$ by using the steepest descent method and the exact steepest descent method. We take the systems $\mathcal{N}(c)$ and $\widetilde{\mathcal{N}(c)}$ with $c = \exp\left(\frac{22}{32}\pi i\right)$, $\exp\left(\frac{26}{32}\pi i\right)$. These examples are exactly those discussed in Sects. 1.5.3 and 3.2.

For $\mathcal{N}(c)$, we have an integral solution

$$\psi(z, \eta) = \int \exp \eta \left(t^4 + c(z - 1)t^2 + zt\right) dt. \tag{A.3.1}$$

The Stokes geometry for $c = \exp\left(\frac{22}{32}\pi i\right)$ is as in Fig. A.5. We investigate a neighborhood of the intersection point $p = 3.159608\cdots + i0.980419\cdots$. As in the right-hand figure, we take 8 values $z = p + 0.03\exp\left(\frac{R\pi i}{180}\right)$ $(R = -20, 3, 27, 47, 154, 184, 207, 225)$. For each value, the steepest descent path of $\mathrm{Re}\left(t^4 + c(z - 1)t^2 + zt\right)$ is given in Fig. A.6.

The change of configuration of steepest descent paths shows that the Stokes curves generated by the double turning point in the z-plane do not cause any Stokes phenomenon.

Meanwhile, the exact steepest descent paths are shown in Figs. A.7 and A.8. In this case, the double turning point can be active (i.e., relevant to Stokes phenomena); The Fourier transform of the equation has a double turning point and a simple pole.

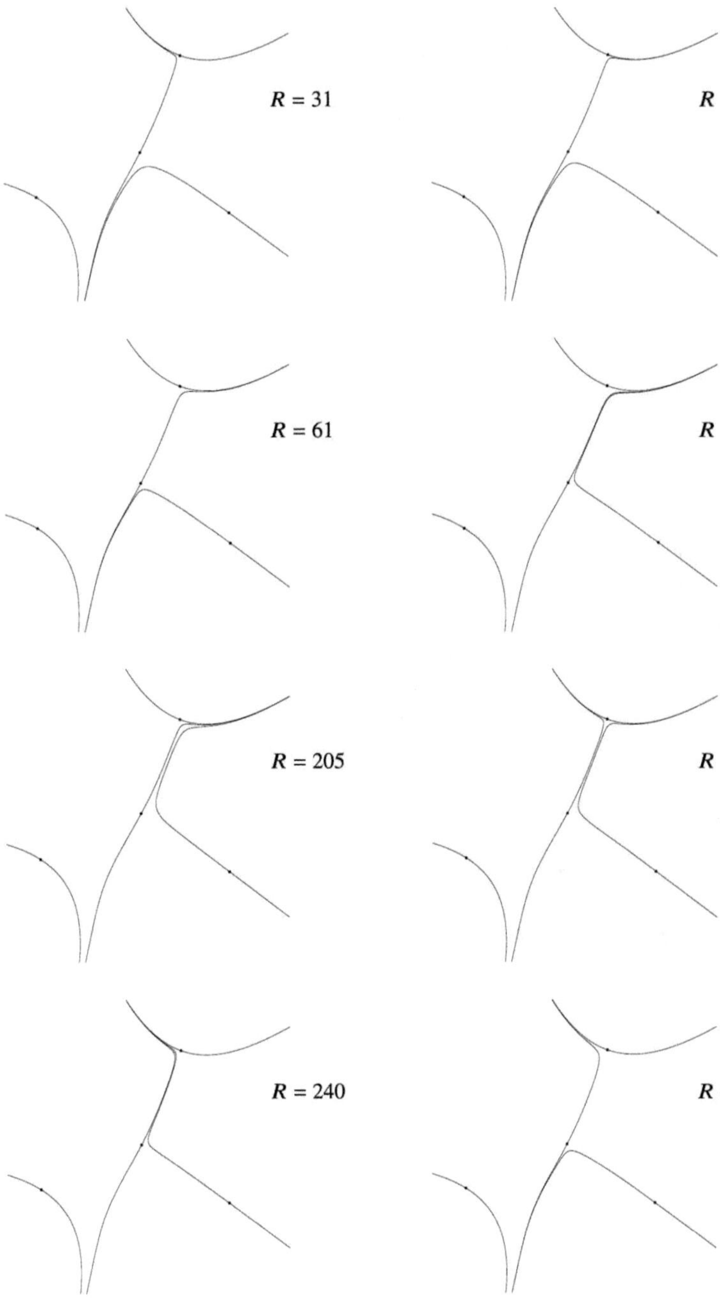

Fig. A.2 The steepest descent paths for $\mathcal{N}_4[x_1 = c_1, x_3 = c_3]$ with $(c_1, c_3) = (\frac{6}{5} + i, -1 - i)$ around the intersection point $p = 4.130956\cdots + i6.724629\cdots$. Small dots denote saddle points. Here 8 figures for $x = p + 0.3\exp\left(\frac{R\pi i}{180}\right)$ $(R = 31, 45, 61, 75, 205, 226, 240, 256)$ are shown

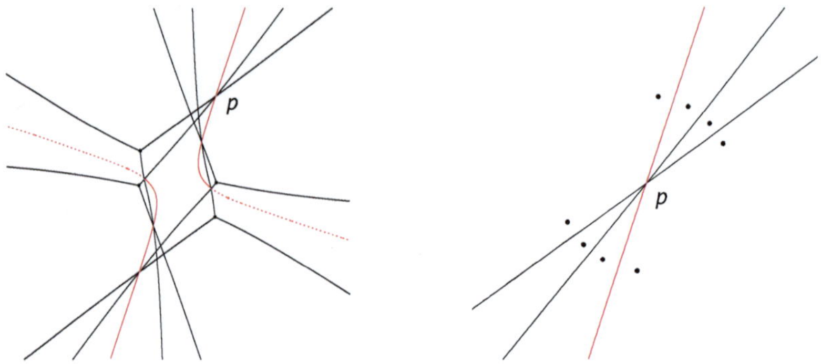

Fig. A.3 Stokes geometry of $\mathcal{N}_4[x_1 = c_1, x_3 = c_3]$ with $(c_1, c_3) = (\frac{6}{5} + i, -i)$. This figure shows the same Stokes geometry as Fig. 1.5, though here inert new Stokes curves are omitted. The right-hand figure is an enlarged version around the intersection point $p = 1.788079\cdots + i4.004568\cdots$

If the original equation is perturbed (i.e., if we consider $\widetilde{\mathcal{N}(c)}$, not $\mathcal{N}(c)$), then the double turning point of the Fourier transformed equation becomes active in general, and a branched steepest descent path causes a Stokes phenomenon on the Stokes curve emanating from the double turning point in the z-plane. We shall explicitly demonstrate this fact.

The explicit form of $\widetilde{\mathcal{N}(c)}$ is given by

$$\eta^{-1}\frac{\partial}{\partial z}\Psi = \left[P_0 + cQ_0 + \eta^{-1}Q_1\right]\Psi \tag{A.3.2}$$

$$= \left[\begin{pmatrix} 0 & 1 & c \\ -\frac{c}{4}z & -\frac{c^2}{2}z + \frac{c(c-z_2)}{2} & 1 \\ -\frac{1}{4}z & -\frac{3c}{4}z + \frac{c-z_2}{2} & -\frac{c^2}{2}z + \frac{c(c-z_2)}{2} \end{pmatrix} + \eta^{-1}\begin{pmatrix} 0 & 0 & 0 \\ 0 & 0 & 0 \\ -\frac{c}{4} & 0 & 0 \end{pmatrix}\right]\Psi.$$

Applying the Fourier transform $\Psi(z, \eta) = \int e^{\eta z\zeta}\hat{\Psi}(\zeta, \eta)\, d\zeta$ and letting $\hat{\Psi} = {}^t(\phi^0, \phi^1, \phi^2) = {}^t(\phi^0, {}^t\Phi)$, we have

$$\eta^{-1}\frac{d}{d\zeta}\Phi = M_1\Phi, \tag{A.3.3}$$

where

$$M_1 = \frac{4}{c^3(1+4c\zeta)}\left[c\zeta(1+2c\zeta-c^3) - \frac{1}{2}(2+6c\zeta-c^3)\begin{pmatrix} 0 & c \\ \zeta & 1 \end{pmatrix} + \eta^{-1}\begin{pmatrix} \frac{c^3}{4\zeta} & \frac{c^4}{4\zeta} \\ \frac{c^3}{2} & \frac{c^4}{2} \end{pmatrix}\right]. \tag{A.3.4}$$

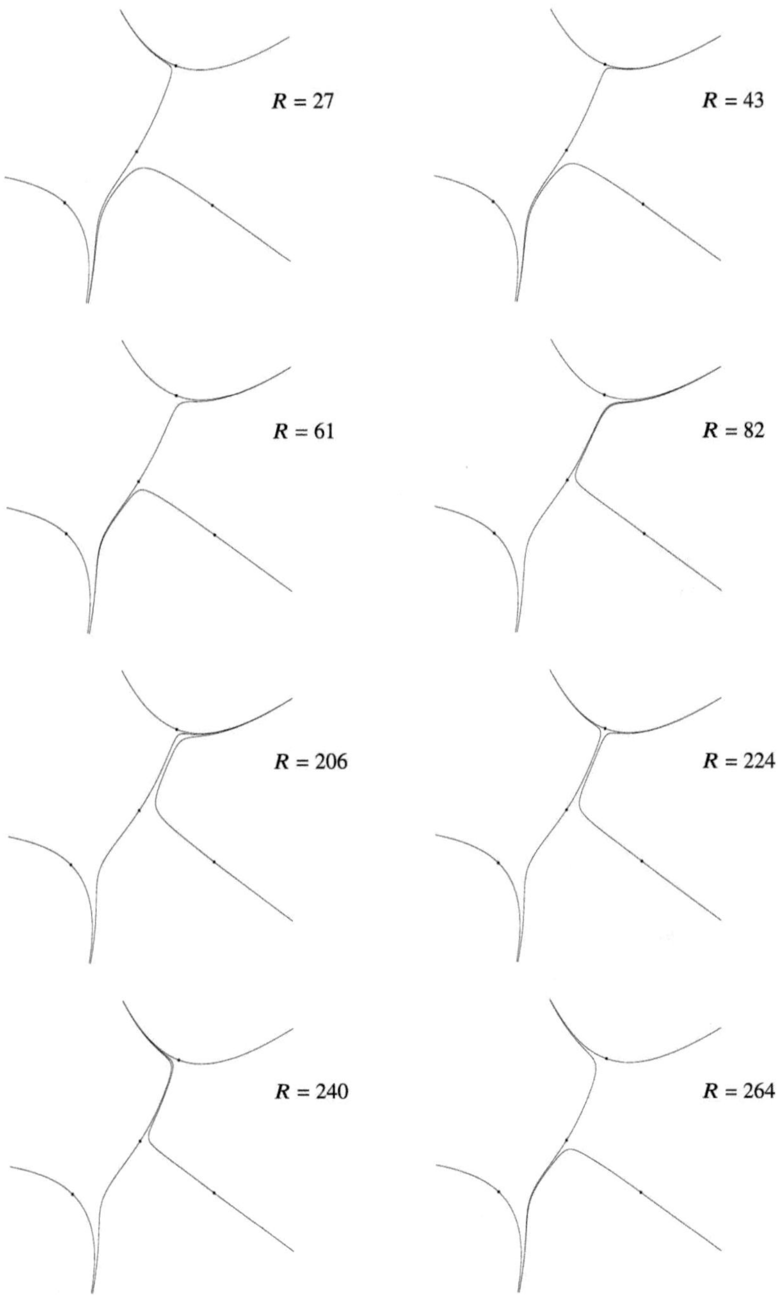

Fig. A.4 The steepest descent paths for $\mathcal{N}_4[x_1 = c_1, x_3 = c_3]$ with $(c_1, c_3) = (\frac{6}{5} + i, -i)$ around the intersection point $p = 1.788079\cdots + i4.004568\cdots$. Small dots denote saddle points. Here 8 figures for $x = p + 0.3\exp\left(\frac{R\pi i}{180}\right)$ $(R = 27, 43, 61, 82, 206, 224, 240, 264)$ are shown

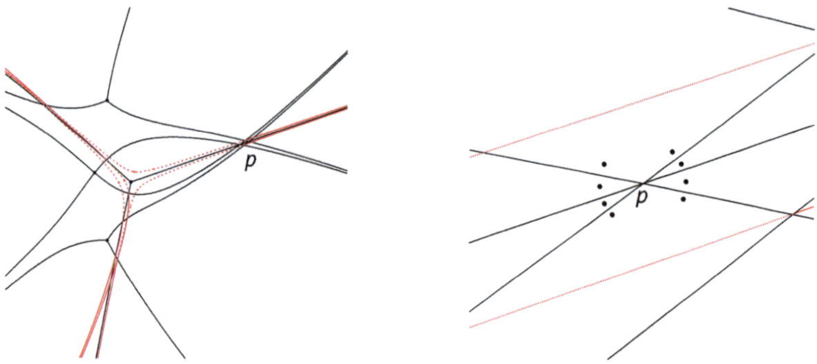

Fig. A.5 Stokes geometry of $\mathcal{N}(c)$ (or $\widetilde{\mathcal{N}(c)}$) with $c = \exp\left(\frac{22}{32}\pi i\right)$. This figure shows the same Stokes geometry as Fig. 1.6, though here some additional new Stokes curves are added to resolve all ordered crossing points. The right-hand figure is an enlarged version around the intersection point $p = 3.159608\cdots + i0.980419\cdots$

This has a double turning point at $\zeta = (c^2/6) - (1/3c)$ and a simple-pole-type turning point at $\zeta = -1/4c$. If the original Eq. (A.3.2) is not perturbed, we have another equation for Φ,

$$\eta^{-1}\frac{\partial}{\partial z_2}\Phi = M_2\Phi, \tag{A.3.5}$$

with

$$M_2 = \left[\frac{c}{2} + \begin{pmatrix} \frac{1}{4\zeta} & \frac{c}{4\zeta} \\ \frac{1}{4} & 0 \end{pmatrix}\right] M_1 - \frac{z_2 - c}{2} - \eta^{-1}\begin{pmatrix} \frac{1}{4\zeta^2} & \frac{c}{4\zeta^2} \\ \frac{1}{4\zeta} & \frac{c}{4\zeta} \end{pmatrix}, \tag{A.3.6}$$

which is compatible with (A.3.3). Then, as is shown in §1.2, the Stokes curves emanating from the double turning point in the ζ-plane actually do not cause any Stokes phenomenon. Otherwise stated, these Stokes curves are inert. Hence they do not cause branching of steepest descent paths. Then, in view of Figs. A.7 and A.8, we see that the Stokes curve in the z-plane which emanates from the double turning point and passes the point p causes no Stokes phenomenon.

On the other hand, if there is perturbation indeed, the Eq. (A.3.5) does not hold any more and the double turning point of (A.3.3) becomes active in general, giving rise to branching of the steepest descent paths as in Figs. A.7 and A.8. Hence the Stokes curve in the z-plane emanating from the double turning point causes a Stokes phenomenon.

In passing, we also calculate the connection coefficient across the Stokes curve emanating from the simple pole $\zeta = -1/4c$, since it is involved for other values of

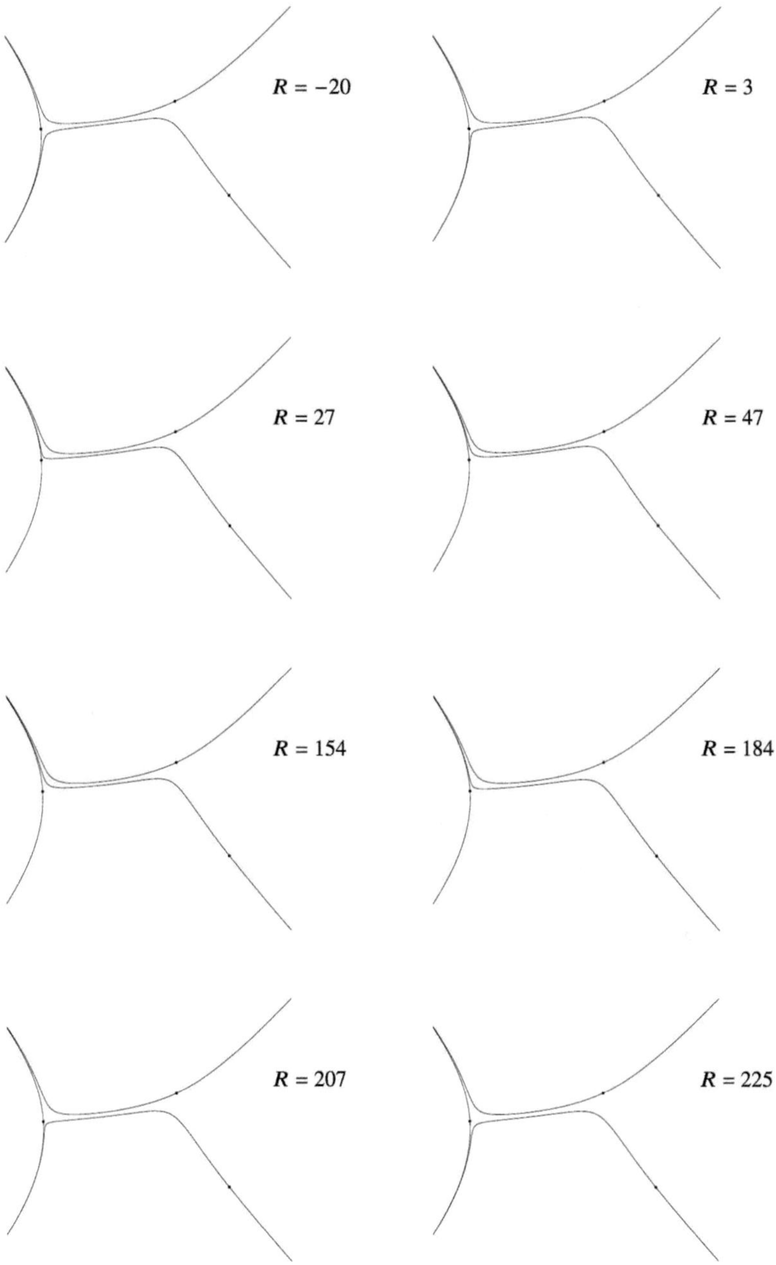

Fig. A.6 The steepest descent paths for $\mathcal{N}(c)$ with $c = \exp\left(\frac{22}{32}\pi i\right)$ around the intersection point $p = 3.159608\cdots + i0.980419\cdots$. Small dots denote saddle points. Here 8 figures for $z = p + 0.03\exp\left(\frac{R\pi i}{180}\right)$ $(R = -20, 3, 27, 47, 154, 184, 207, 225)$ are shown

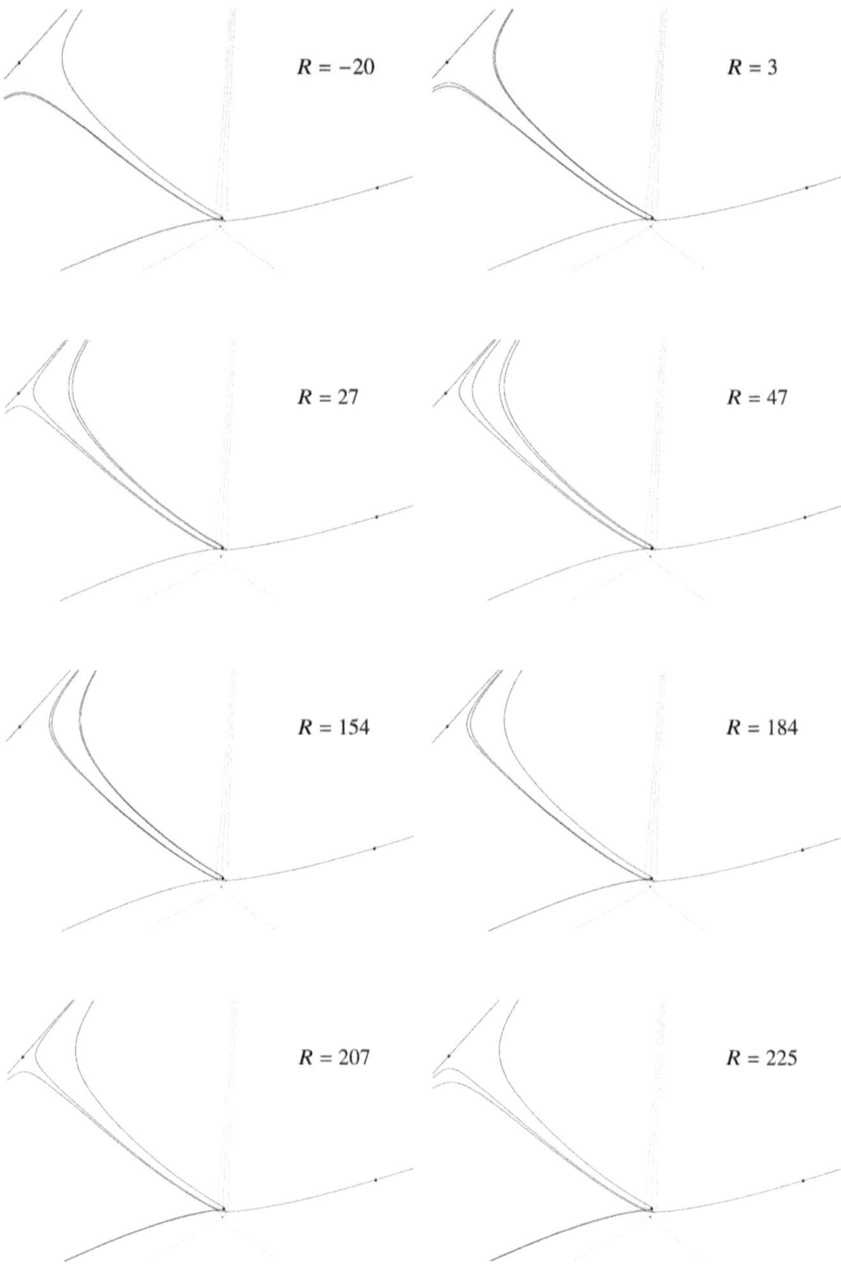

Fig. A.7 The exact steepest descent paths for $\mathcal{N}(c)$ with $c = \exp\left(\frac{22}{32}\pi i\right)$ around the intersection point $p = 3.159608\cdots + i0.980419\cdots$. Small dots denote saddle points. Blue curves are Stokes curves of the Fourier transformed equation. Here 8 figures for $z = p + 0.03\exp\left(\frac{R\pi i}{180}\right)$ ($R = -20, 3, 27, 47, 154, 184, 207, 225$) are shown

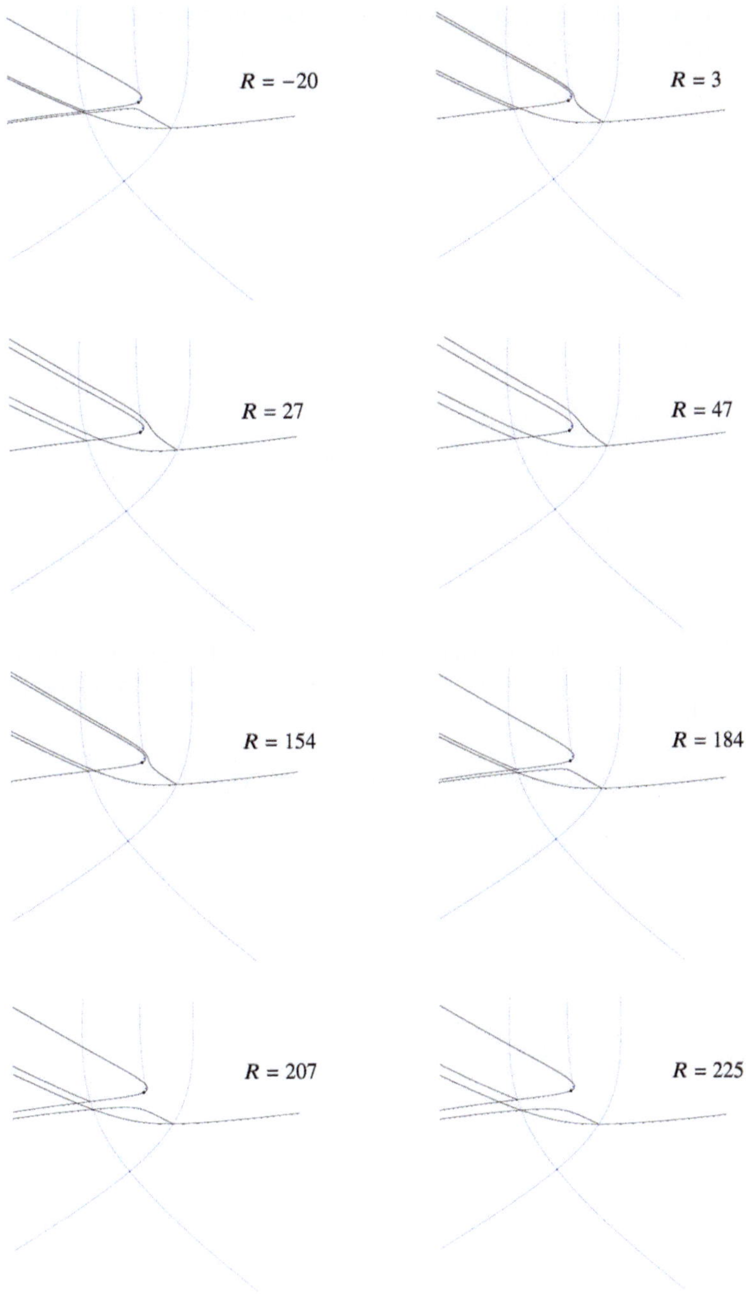

Fig. A.8 The exact steepest descent paths for $\mathcal{N}(c)$ with $c = \exp\left(\frac{22}{32}\pi i\right)$ around the intersection point $p = 3.159608\cdots + i0.980419\cdots$ (enlarged). Here 8 figures for $z = p + 0.03\exp\left(\frac{R\pi i}{180}\right)$ $(R = -20, 3, 27, 47, 154, 184, 207, 225)$ are shown

z (cf. [HiKST, Figure 3.4]). Let f denote the first component of Φ. Then we have a single equation

$$\left(\frac{d^2}{d\zeta^2} - \eta P(\zeta,\eta)\frac{d}{d\zeta} - \eta^2 Q(\zeta,\eta)\right) f = 0 \tag{A.3.7}$$

with

$$P(\zeta,\eta) = \frac{4}{c^3}\left(1 + \zeta c - \frac{c^3}{2}\right) + \eta^{-1}\left(\frac{\frac{1}{\zeta} - 2c}{1 + 4\zeta c} + \frac{6c + \eta^{-1}\frac{c^3}{2\zeta^2}}{2 + 6\zeta c - c^3 - \eta^{-1}\frac{c^3}{2\zeta}}\right) \tag{A.3.8}$$

$$Q(\zeta,\eta) = \frac{q_0(\zeta) + \eta^{-1}q_1(\zeta) + \eta^{-2}q_2(\zeta)}{c^3\zeta(1 + 4\zeta c)(2 + 6\zeta c - c^3 - \eta^{-1}\frac{c^3}{2\zeta})}, \tag{A.3.9}$$

where q_0, q_1 and q_2 are some polynomials. If we write

$$P(\zeta,\eta) = P_0(\zeta) + \eta^{-1}P_1(\zeta) + \eta^{-2}P_2(\zeta) + \cdots,$$

then only P_1 has a simple pole. Hence this is a simple-pole-type operator studied in [KKoT] and we can compute the Stokes multiplier by using the result of [Ko]. In fact, if we let

$$Q(\zeta,\eta) = Q_0(\zeta) + \eta^{-1}Q_1(\zeta) + \eta^{-2}Q_2(\zeta) + \cdots,$$

then the Stokes multiplier is given by

$$\pm 2i\cos(\pi\sqrt{1 + 4\lambda}), \tag{A.3.10}$$

where

$$\lambda = \frac{1}{4}\left(\operatorname*{Res}_{\zeta=-1/4c} P_1(\zeta)\right)^2 + \frac{1}{2}\operatorname*{Res}_{\zeta=-1/4c} P_1(\zeta) \tag{A.3.11}$$

$$= \frac{1}{4}\left(-\frac{3}{2}\right)^2 + \frac{1}{2}\left(-\frac{3}{2}\right) = -\frac{3}{16}. \tag{A.3.12}$$

Thus, we see that the simple pole in the ζ-plane is inert when the system $\mathcal{N}(c)$ is not perturbed (though it can be active when the system is perturbed).

Then we consider $\widetilde{\mathcal{N}(c)}$ with a different value of parameter $c = \exp\left(\frac{26}{32}\pi i\right)$. The Stokes geometry is as in Fig. A.9. Observe that, as c changes from $\exp\left(\frac{22}{32}\pi i\right)$ to $\exp\left(\frac{26}{32}\pi i\right)$, a Stokes curve which emanates from the double turning point and passes the intersection point p is replaced by a new Stokes curve emanating from a virtual turning point. Again, we take 8 values $z = p + 0.03\exp\left(\frac{R\pi i}{180}\right)$

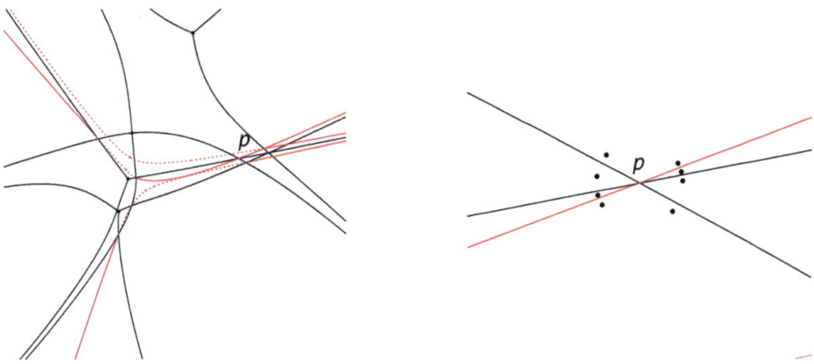

Fig. A.9 Stokes geometry of $\mathcal{N}(c)$ (or $\widetilde{\mathcal{N}(c)}$) with $c = \exp\left(\frac{26}{32}\pi i\right)$. This figure shows the same Stokes geometry as Fig. 1.8, though here some additional new Stokes curves are added to resolve all ordered crossing points. The right-hand figure is an enlarged version around the intersection point $p = 3.212067\cdots + i0.646373\cdots$

($p = 3.212067\cdots + i0.646373\cdots$, $R = -40, 3, 15, 27, 140, 171, 196, 210$). For each value, the steepest descent path of $\mathrm{Re}\left(t^4 + c(z-1)t^2 + zt\right)$ is given in Fig. A.10. The change of configuration of steepest descent paths shows that the Stokes curve emanating from the double turning point does not cause any Stokes phenomenon. Meanwhile, the exact steepest descent path is as in Figs. A.11 and A.12. Note that the double turning point in the ζ-plane is again inert when we consider the non-perturbed system $\mathcal{N}(c)$. Thus the change of configuration of steepest descent paths and exact steepest descent paths is consistent with the Stokes geometry in the z-plane, despite the fact that some ordinary Stokes curves and new Stokes curves are interchanged as the value of parameter c varies.

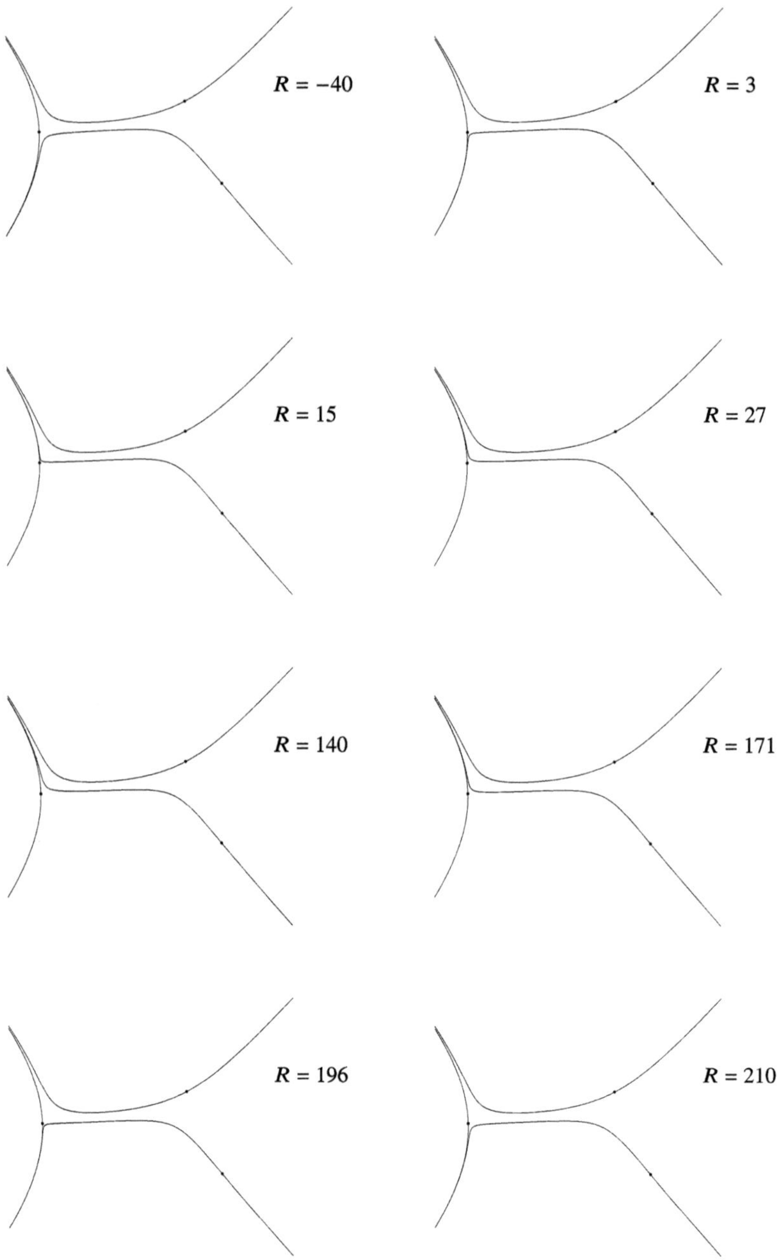

Fig. A.10 The steepest descent paths for $\mathcal{N}(c)$ with $c = \exp\left(\frac{26}{32}\pi i\right)$ around the intersection point $p = 3.212067\cdots + i0.646373\cdots$. Small dots denote saddle points. Here again 8 figures for $z = p + 0.03\exp\left(\frac{R\pi i}{180}\right)$ ($R = -40, 3, 15, 27, 140, 171, 196, 210$) are shown

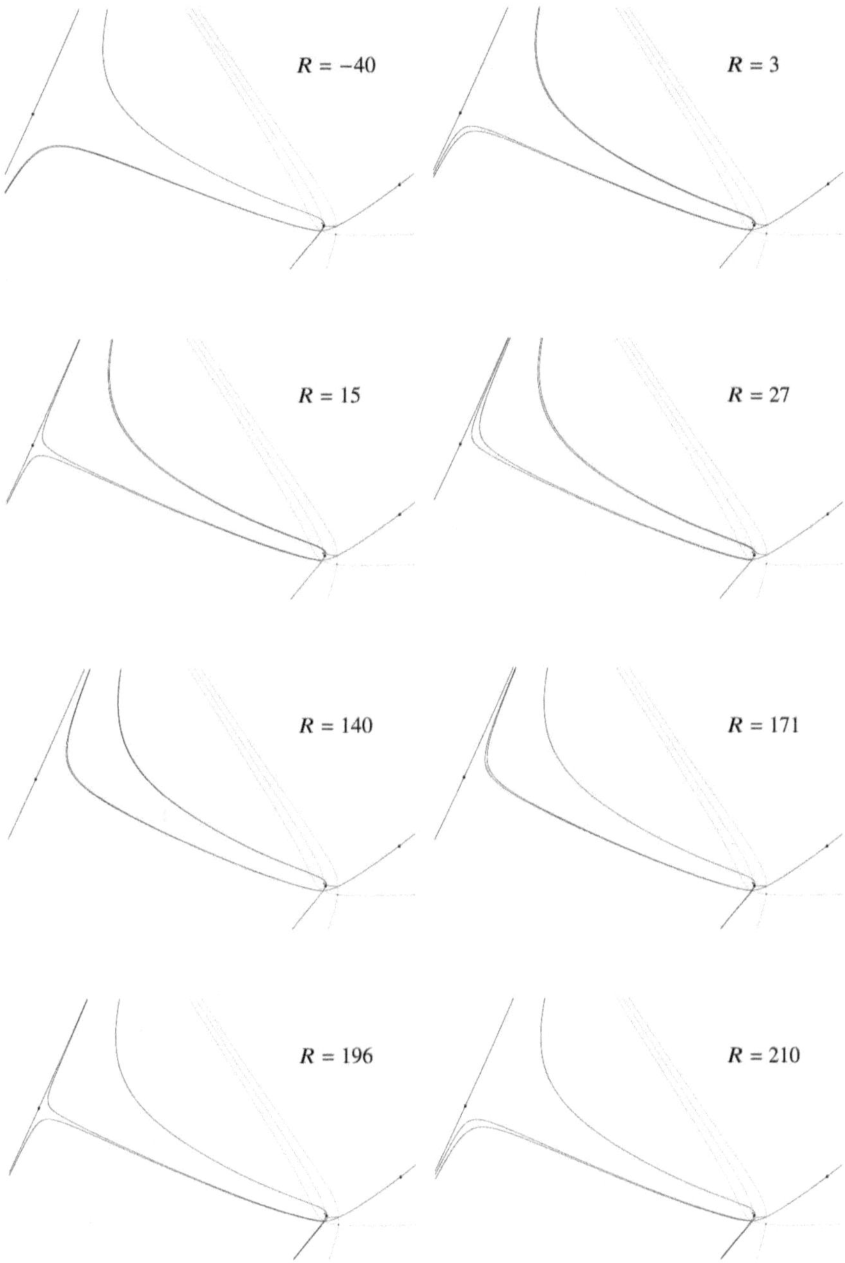

Fig. A.11 The exact steepest descent paths for $\mathcal{N}(c)$ with $c = \exp\left(\frac{26}{32}\pi i\right)$ around the intersection point $p = 3.212067\cdots + i0.646373\cdots$. Small dots denote saddle points. Blue curves are Stokes curves of the Fourier transformed equation. Again 8 figures for $z = p + 0.03\exp\left(\frac{R\pi i}{180}\right)$ ($R = -40, 3, 15, 27, 140, 171, 196, 210$) are shown

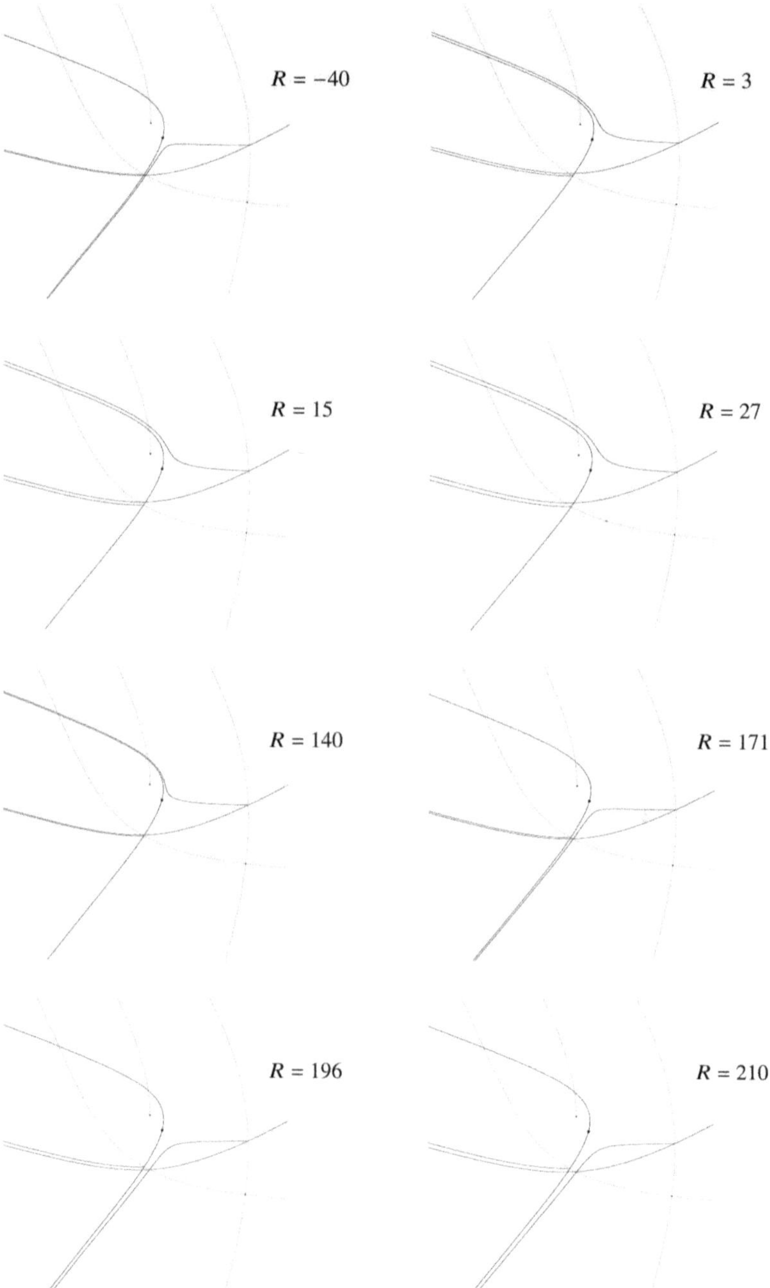

Fig. A.12 The exact steepest descent paths for $\mathcal{N}(c)$ with $c = \exp\left(\frac{26}{32}\pi i\right)$ around the intersection point $p = 3.212067\cdots + i0.646373\cdots$ (enlarged). Again 8 figures for $z = p + 0.03\exp\left(\frac{R\pi i}{180}\right)$ $(R = -40, 3, 15, 27, 140, 171, 196, 210)$ are shown

References

[VTP] N. Honda, T. Kawai, Y. Takei, *Virtual Turning Points*. Springer Briefs in Mathematical Physics, vol. 4 (Springer-Verlag, Berlin, 2015)

[A] T. Aoki, Toward the exact WKB analysis of holonomic systems. RIMS Kôkyûroku **1433**, 1–8 (2005) (in Japanese)

[AHKKoNSST] T. Aoki, N. Honda, T. Kawai, T. Koike, Y. Nishikawa, S. Sasaki, A. Shudo, Y. Takei, Virtual turning points — A gift of microlocal analysis to the exact WKB analysis. *Algebraic Analysis of Differential Equations* (Springer, Berlin, 2008), pp. 29–43

[AKKoST] T. Aoki, T. Kawai, T. Koike, S. Sasaki, Y. Takei, Virtual turning points and isomonodromic deformations — On the observation of S. Sasaki for the creation of new Stokes curves of Noumi-Yamada systems —. RIMS Kôkyûroku **1493**, 191–211 (2006)

[AKSST] T. Aoki, T. Kawai, S. Sasaki, A. Shudo, Y. Takei, Virtual turning points and bifurcation of Stokes curves. J. Phys. **A 38**, 3317–3336 (2005)

[AKT1] T. Aoki, T. Kawai, Y. Takei, New turning points in the exact WKB analysis for higher-order ordinary differential equations. *Analyse algébrique des Perturbations Singulières, I* (Hermann, Paris, 1994), pp. 69–84

[AKT2] T. Aoki, T. Kawai, Y. Takei, On the exact steepest descent method: A new method for the description of Stokes curves. J. Math. Phys. **42**, 3691–3713 (2001)

[AKT3] T. Aoki, T. Kawai, Y. Takei, *The Bender-Wu Analysis and the Voros Theory. II*. Advanced Studies in Pure Mathematics, vol. 54 (Mathematical Society of Japan, Tokyo, 2009), pp. 19–94

[BNR] H.L. Berk, W.M. Nevins, K.V. Roberts, New Stokes' line in WKB theory. J. Math. Phys. **23**, 988–1002 (1982)

[F] M.V. Fedoryuk, Asymptotic methods in analysis. Encyclopedia Math. Sci. **13**, 83–191 (1989)

[Hi1] S. Hirose, On the Stokes geometry for the Pearcey system and the (1,4) hypergeometric system. RIMS Kôkyûroku Bessatsu **B40**, 243–292 (2013)

[Hi2] S. Hirose, On a non-hereditary turning point of a tangential system of the Pearcey system. RIMS Kôkyûroku Bessatsu **B75**, 138–150 (2019)

[HiKT] S. Hirose, T. Kawai, Y. Takei, Pearcey system re-examined from the viewpoint of s-virtual turning points and non-hereditary turning points. RIMS Kôkyûroku Bessatsu **B75**, 151–175 (2019)

[HiKST] S. Hirose, T. Kawai, S. Sasaki, Y. Takei, On the Stokes geometry of perturbed tangential Pearcey systems. Publ. RIMS, Kyoto Univ. **57**, 727–754 (2021)

S. Hirose et al., *Virtual Turning Points II*, SpringerBriefs in Mathematical Physics 52, https://doi.org/10.1007/978-981-95-2817-2

[H] N. Honda, The geometric structure of a virtual turning point and the model of the Stokes geometry. RIMS Kôkyûroku Bessatsu **B10**, 63–117 (2009)

[KKO] M. Kashiwara, T. Kawai, T. Oshima, Structure of cohomology groups whose coefficients are microfunction solution sheaves of systems of pseudo-differential equations with multiple characteristics, I. Proc. Jpn. Acad. **49**, 420–425 (1974)

[KKoT] T. Kawai, T. Koike, Y. Takei, On the exact WKB analysis of higher order simple-pole type operators. Adv. Math. **228**, 63–96 (2011)

[KT] T. Kawai, Y. Takei, *Algebraic Analysis of Singular Perturbation Theory* (American Mathematical Society, 2005) (Translated from the Japanese edition published by Iwanami in 1998)

[Ko] T. Koike, On the exact WKB analysis of second order linear ordinary differential equations with simple poles. Publ. RIMS Kyoto Univ. **36**, 297–319 (2000)

[KoS] T. Koike, R. Schäfke, On the Borel summability of WKB solutions of Schrödinger equations with polynomial potentials and its application. In preparation

[NY] M. Noumi, Y. Yamada, Symmetry in painlevé equations. *Toward the Exact WKB Analysis of Differential Equations. Linear or Non-Linear* (Kyoto University Press, Kyoto, 2000), pp. 245–260

[OK] K. Okamoto, H. Kimura, On particular solutions of the Garnier systems and the hypergeometric functions of several variables. Q. J. Math. **37**, 61–80 (1986)

[Sa1] S. Sasaki, On parametric Stokes phenomena of higher order linear ODEs. RIMS Kôkyûroku Bessatsu **B61**, 163–180 (2017)

[Sa2] S. Sasaki, Borel summability of WKB-theoretic transformation near a double turning point. To appear in RIMS Kôkyûroku Bessatsu

[S1] S. Sasaki, On the role of virtual turning points in the deformation of higher order linear differential equations. RIMS Kôkyûroku **1433**, 27–64 (2005)

[S2] S. Sasaki, On the role of virtual turning points in the deformation of higher order linear differential equations, II. RIMS Kôkyûroku **1433**, 65–109 (2005)

[SKK] M. Sato, T. Kawai, M. Kashiwara, *Microfunctions and Pseudo-differential Equations*. Lecture Notes in Mathematics, vol. 287 (Springer, Berlin, 1973), pp. 265–529

[T1] Y. Takei, Integral representation for ordinary differential equations of Laplace type and exact WKB analysis. RIMS Kôkyûroku **1168**, 80–92 (2000)

[T2] Y. Takei, An explicit description of the connection formula for the first Painlevé equation. *Toward the Exact WKB Analysis of Differential Equations. Linear or Non-Linear* (Kyoto University Press, Kyoto, 2000), pp. 271–296

[T3] Y. Takei, On a double turning point problem for systems of linear ordinary differential equations. Preprint

[T4] Y. Takei, Toward the exact WKB analysis for higher-order Painlevé equations – The case of Noumi-Yamada Systems. Publ. RIMS Kyoto Univ. **40**, 709–730 (2004)

[V] A. Voros, The return of the quartic oscillator. The complex WKB methods. Ann. Inst. H. Poincaré **39**, 211–338 (1983)

Index

S. Hirose et al., *Virtual Turning Points II*, SpringerBriefs in Mathematical Physics 52, https://doi.org/10.1007/978-981-95-2817-2

FSC
www.fsc.org
MIX
Papier aus verantwortungsvollen Quellen
Paper from responsible sources
FSC® C105338

If you have any concerns about our products,
you can contact us on
ProductSafety@springernature.com

In case Publisher is established outside the EU,
the EU authorized representative is:
Springer Nature Customer Service Center GmbH
Europaplatz 3, 69115 Heidelberg, Germany

Printed by Libri Plureos GmbH
in Hamburg, Germany